D0897313

Open: The Philosophy and Practices that are Revolutionizing Education and Science

Edited by
Rajiv S. Jhangiani and
Robert Biswas-Diener

]u[

ubiquity press
London

Published by
Ubiquity Press Ltd.
6 Windmill Street
London W1T 2JB
www.ubiquitypress.com

Text © The Authors 2017

First published 2017

Cover design by Amber MacKay
Images used in the cover design are licensed under CC0 Public Domain.
Main cover image: Pexels/Pixabay.com
Background cover image: lalesh aldarwish/Pexels.com

Printed in the UK by Lightning Source Ltd.
Print and digital versions typeset by Siliconchips Services Ltd.

ISBN (Paperback): 978-1-911529-00-2
ISBN (PDF): 978-1-911529-01-9
ISBN (EPUB): 978-1-911529-02-6
ISBN (Mobi): 978-1-911529-03-3

DOI: https://doi.org/10.5334/bbc

The full text of this book has been peer-reviewed to ensure high academic standards. For full review policies, see http://www.ubiquitypress.com/

Suggested citation:
Jhangiani, R S and Biswas-Diener, R 2017 *Open: The Philosophy and Practices that are Revolutionizing Education and Science*. London: Ubiquity Press. DOI: https://doi.org/10.5334/bbc. License: CC-BY 4.0

To read the free, open access version of this book online, visit https://doi.org/10.5334/bbc or scan this QR code with your mobile device:

Table of Contents

Acknowledgements   vi

Contributors   vii

Introduction   1

Introduction to Open (Robert Biswas-Diener and
Rajiv S. Jhangiani)   3

A Brief History of Open Educational Resources (T. J. Bliss and
M. Smith)   9

Open Licensing and Open Education Licensing Policy
(Cable Green)   29

Openness and the Transformation of Education and Schooling
(William G. Huitt and David M. Monetti)   43

What Can OER Do for Me? Evaluating the Claims for OER
(Martin Weller, Beatriz de los Arcos, Rob Farrow,
Rebecca Pitt and Patrick McAndrew)   67

Are OE Resources High Quality? (Regan A. R. Gurung)   79

Open Practices   87

Opening Science (Brian A. Nosek)   89

Open Course Development at the OERu (Wayne Mackintosh)   101

From OER to Open Pedagogy: Harnessing the Power of Open
(Robin DeRosa and Scott Robinson)   115

Opening Up Higher Education with Screencasts
(David B. Miller and Addison Zhao)   125

Librarians in the Pursuit of Open Practices (Quill West)   139

A Library Viewpoint: Exploring
Open Educational Practices (Anita Walz)   147

How to Open an Academic Department (Farhad Dastur)   163

Case Studies   **179**

The International Journal of Wellbeing: An Open Access
Success Story (Dan Weijers and Aaron Jarden)   181

Iterating Toward Openness: Lessons Learned on a
Personal Journey (David Wiley)   195

Open-Source for Educational Materials Making Textbooks
Cheaper and Better (Ed Diener, Carol Diener and
Robert Biswas-Diener)   209

Free is Not Enough (Richard Baraniuk, Nicole Finkbeiner,
David Harris, Dani Nicholson and Daniel Williamson)   219

The BC Open Textbook Project (Mary Burgess)   227

TeachPsychScience.org: Sharing to Improve the Teaching of
Research Methods (David B. Strohmetz, Natalie J. Ciarocco and
Gary W. Lewandowski, Jr.)   237

DIY Open Pedagogy: Freely Sharing Teaching Resources in
Psychology (Jessica Hartnett)   245

Conclusion   255

You Can't Sell Free, and Other OER Problems
(Robert Biswas-Diener)   257

Open as Default: The Future of Education and
Scholarship (Rajiv S. Jhangiani)   267

Index   281

Acknowledgements

No book—whether edited or single authored—is truly an individual effort. We are grateful to those who helped us improve the quality of this volume. First, we would like to thank Nadia Lyubchik for her support in preparing the manuscript for publication. Just as a script supervisor ensures the continuity of a feature film, Nadia kept track of all of the references, formatting, correspondence and countless other details necessary for the publication of this book. Our deepest thanks. We would also like to thank Peter Lindberg at Noba, who gave valuable feedback on a number of aspects of this project as well as specific chapters. We would also like to thank Cathy Casserly at the Carnegie Foundation for her input. Finally, we are grateful to the contributing authors for their eager participation in this project. The breadth and depth of their collective insights and experiences have made this volume everything an editor could hope for. This book was produced with the support of a grant from the Association for Psychological Science and we gratefully acknowledge their role. We would also like to thank David Ernst and George Veletsianos for providing informative peer review feedback and helping us to develop the book.

Contributors

Richard Baraniuk – Rice University, OpenStax College
Robert Biswas-Diener – Noba Project
T. J. Bliss – Hewlett Foundation
Mary Burgess – BCcampus
Natalie Ciarocco – Monmouth University
Farhad Dastur – Kwantlen Polytechnic University
Beatriz de los Arcos – The Open University
Robin DeRosa – Plymouth State University
Carol Diener – University of Illinois
Ed Diener – University of Virginia
Rob Farrow – The Open University
Nicole Finkbeiner – OpenStax College
Cable Green – Creative Commons
Regan Gurung – University of Wisconsin-Green Bay
David Harris – OpenStax College
Jessica Hartnett – Gannon University
William Huitt – Valdosta State University & Capella University
Aaron Jarden – Auckland University of Technology
Rajiv S. Jhangiani – Kwantlen Polytechnic University
Gary Lewandowski – Monmouth University
Wayne Mackintosh – Otago Polytechnic & OER Foundation
Patrick McAndrew – The Open University
David Miller – University of Connecticut
David Monetti – Valdosta State University
Danielle Nicholoson – OpenStax College
Brian Nosek – University of Virginia & Center for Open Science
Rebecca Pitt – The Open University
Scott Robinson – Portland State University
Mike Smith – Hewlett Foundation
David Strohmetz – Monmouth University
Anita Walz – Virginia Polytechnic Institute & State University
Daniel Weijers – University of Waikato

Martin Weller – The Open University
Quill West – Pierce College
David Wiley – Lumen Learning
Daniel Williamson – OpenStax College
Addison Zhao – University of Connecticut

Introduction

Introduction to Open

Robert Biswas-Diener* and Rajiv S. Jhangiani[†]
*Noba Project, robert@nobaproject.com
[†]Kwantlen Polytechnic University

The creation and spread of knowledge has always redefined the possibilities of the human experience. Among all the inventions of humans—water treatment, bows and arrows, space travel—formal education is, perhaps, the most powerful. Whether in the form of trade apprenticeships, religious schools, or modern universities, education is the principal way in which we pass skills and knowledge from one person, and even one generation, to another. At the heart of education lies an inquiry and understanding of how it is that we come to know. In modern times this includes the development and application of the scientific method, which has become vital to the creation and dissemination of knowledge. As education becomes more accessible, affordable, and flexible, knowledge and skills act, increasingly, as tools for the empowerment for the world's poor. In fact, studies point to education as being associated with better health, improved well-being, and increased economic empowerment.[1]

This notion that education can unlock a person's potential is relatively non-controversial. More controversial is the related notion that education should be made available to all. Indeed, despite the fact that it is specifically mentioned in Article 26 of the United Nations Universal Declaration of Human rights,[2] this 'education as a fundamental human right' idea is a relatively recent phenomenon. Historically, learning was an activity reserved for privileged citizens: the priestly class, the wealthy, men. In *The Theory of the Leisure Class*, Veblen

How to cite this book chapter:
Biswas-Diener, R and Jhangiani, R S. 2017. Introduction to Open. In: Jhangiani, R S and Biswas-Diener, R. (eds.) *Open: The Philosophy and Practices that are Revolutionizing Education and Science.* Pp. 3–7. London: Ubiquity Press. DOI: https://doi.org/10.5334/bbc.a. License: CC-BY 4.0

(1899/2009) argues that the 'learned class' has therefore long been associated with the aspects of higher education that have no immediate industrial use, such as the study of the classics or music. However, over the last couple of decades advances in technology have enabled the marginal cost of education, such as the sharing of resources, to approach zero.[3] This means that barriers to education are now being removed for a broader swath of humanity than at any time in history. What is more, that all people should have equal access to not only job skills-related education but also a liberal arts education is a case that can be made more strongly on a human rights platform than on an economic empowerment platform. The real tragedy of poverty is not that the poor need more opportunities to be factory foremen, office managers, or stock traders. The real tragedy is the loss of every scientific discovery, artistic work, invention, new business, and cultivated mind because of lack of opportunity according to random chance at birth.

Unfortunately, educational inequality abounds. Braun (2014) states that as many as 57 million school-aged children worldwide do not attend primary school. In 1999 the ambitious *Voices of the Poor* project was launched in order to conduct field interviews with tens of thousands of poor people from around the globe.[4] With regard to education, poor people identified the strain of costly and distant schools, lack of quality education, feelings of low self-worth, and competition with child labor as widespread educational problems. In his study of poverty in Kolkata (Calcutta), India Thomas (1997) echoes this last point:

'Among the poor, children often have to start work at the age of eight and so are unable to complete their primary education. This means that the entire blue-collar labor force will continue to be functionally illiterate for another generation' (Thomas, 1997: 117).

Educational inequalities are as much a reality in affluent and industrialized societies as they are in developing economies. In countries as diverse as New Zealand, Canada, and the United Kingdom, the histories of colonization and immigration have been associated with disparities in access to high quality education. Kozol (1992) points to racial segregation as a primary source of what he calls 'savage inequalities.' He traces systematic differences in per pupil expenditure, funding structure, and facilities between affluent and poor minority school districts in the United States. This trend endures in the United States to the present: high schoolers taking advanced placement or international baccalaureate courses consistently outperform their less advanced counterparts on various measures of academic achievement and poor students are underrepresented among the educational elect.[5] In just the first decade of this century, 2.4 million American students either did not attend, or could not complete, college because of the cost barrier.[6]

The open education movement offers one possible, partial remedy to educational inequality. The most obvious benefit of open education is in its low cost. The word 'open,' in this sense, means 'allowing access to' although it is also often equated with 'free of cost.' In fact, most open education resources are freely available and even in cases where they are low cost, they still help to drive the market toward a lower price point. By removing or substantially reducing the expense normally associated with software, textbooks, and course fees, education becomes more accessible to more people. The open education movement can also help raise the quality of education for all students because instructors are better able to share and build on one another's pedagogical innovations. It is here, in the second sense of 'open,' meaning customizable by and shareable among instructors, that we have the potential to design more engaging, locally relevant, interactive, and effective teaching resources.

It is for precisely these reasons that open education often seems like a crusade. It is a values-based and mission-driven movement every bit as much as it is practical and technological. The voices of open advocates and champions are often impassioned in the way typical of people who are in the throes of rapid and successful social change. The editors of this volume are no different. We prize education for what it can unlock and experience a great deal of purpose in our role as instructors. As we became acquainted with open practices we fell in love. It was when this passion rose to a boiling point that we decided to create this volume.

This book is intended to share the principal voices, motivations, and practices of the open movement. Most of our contributors work within academia (these include faculty, librarians, and learning technologists), while others work to support the movement from within the private and non-profit sectors. They all care deeply about scientific progress, the democratization of education, pedagogical innovation, and the many ways in which these goals support one another. The practices they discuss encompass a broad range, including the creation, adaptation, and adoption of open educational resources, open pedagogy, open course development, open science, and open access. Despite this heterogeneity, they all wage parallel battles for access and progress and against territoriality and tradition (including traditional notions of prestige).

You will notice several themes emerge across the chapters in this volume. In addition to the obvious focus on access, these include transparency, flexibility, credibility, and creativity. Access concerns scholarly publications, data, required course materials, and, of course, tertiary education itself. Transparency is manifested in the pre-registration of research hypotheses, course development on the open web, and faculty reviews of open textbooks. Flexibility includes the contextualization of open educational resources, hybrid delivery models, and flexible learning pathways built across an international network. Credibility is seen in the support of Open Educational Resources (OER) development by professional bodies, leading scholars serving on editorial boards of open access journals, and research on the impact of open textbooks on learning outcomes. And creativity is exhibited by student–creators of OER, collaborative efforts to develop ancillary resources, and the development of licenses, organizations, repositories, and other infrastructure to support the open movement.

We encouraged our contributors to share their personal journey along with their hard-earned insights. Of course, the personal narratives of Wiley, Green, Weller, Mackintosh, Bliss, and Smith chronicle the evolution of the open education movement itself. Whereas some chapters tell the story of large projects funded by government (Burgess) and philanthropic organizations (Baraniuk et al.), others (Hartnett; Diener et al.) reveal initiatives by individual rebel-pioneers. Several chapters double as practical how-to guides, whether for starting an open access journal (Weijers & Jarden), developing a platform to support collaborative scientific inquiry (Nosek), freely sharing teaching and learning resources (Strohmetz, Ciarocco, & Lewandowski; Miller & Zhao), redesigning course assignments to allow students to practice public scholarship (DeRosa & Robinson), advocating for change within an academic department (Dastur), or supporting open access and open education from a university library (Walz; West). As much as this book reflects on the journey of the open movement up to this point, it also looks to the future – including the challenges we must navigate and the opportunities we must seize – if open is to become the default practice.

Initially, we conceived of this book as being primarily about open practices and resources as applied to our own discipline, psychology. Both of us work as researchers and instructors of psychology and we felt the naturalness of sharing open with our colleagues. As the chapters from our contributors arrived, however, we realized that the scope of these contributions were far broader than just our corner of the social sciences. Quite simply, the expert commentary on the history, current trends, and future of open education and science were too good to confine to psychology. Even so, readers should know that that initial framework, open psychology, still casts its shadow over this volume. You can see, for instance, a conspicuous number of contributions from psychologists (about half the chapters). You will also note that many broad concepts in open education such as open pedagogy, open textbooks, and open departments are illustrated in this book with examples drawn from psychology. In the opinions of the editors these should be treated as examples only and we hope that readers will feel empowered to modify these ideas in ways that fit their own disciplines.

In closing, we pose a direct call to readers. Open education, open science, open access, and open pedagogy are new phenomena. They are imperfect and many challenges remain to be overcome. However, as the open movement matures and gains momentum, and as the questions it poses grow increasingly nuanced, the boundaries of the movement continue to expand. The open movement represents both an optimistic promise for the future as well as a myriad of practical tools and strategies for the present. Although it is true that we hope to convince you of the merits of open, we do not demand that you 'convert.' Open is a gift on offer. Like any gift, it is up to you whether you think it is worthwhile to accept it. We only ask that you consider.

Notes

[1] Orr, Rimini & van Damme, 2015.
[2] UN General Assembly, 1948.
[3] Rifkin, 2014.
[4] Narayan et al., 2000.
[5] Godsey, 2015.
[6] ACFSA, 2006.

References

Advisory Committee on Student Financial Assistance (ACFSA). (2006). *Mortgaging our future: How financial barriers to college undercut America's global competitiveness.* Retrieved from http://files.eric.ed.gov/fulltext/ED529499.pdf

Braun, A. (2014). *The promise of a pencil.* New York, NY: Scribner.

Godsey, M. (2015, June 15). The inequality in public schools. *The Atlantic.* Retrieved from http://www.theatlantic.com/education/archive/2015/06/inequality-public-schools/395876/

Kozol, J. (1992). *Savage inequalities: Children in America's schools.* New York, NY: Broadway Paperbacks.

Narayan, D., Chambers, R., Shah, M. K., & Petesch, P. (2000). *Voices of the poor: Crying out for change.* New York, NY: Oxford University Press.

Orr, D., Rimini, M., & van Damme, D. (2015). *Open educational resources: A catalyst for innovation.* Paris, France: OECD Publishing.

Rifkin, J. (2014). *The zero marginal cost society: The internet of things, the collaborative commons, and the eclipse of capitalism.* New York, NY: St. Marten's Press.

Thomas, F. C. (1997). *Calcutta: The human face of poverty.* London, UK: Penguin Books.

UN General Assembly. (1948). *Universal declaration of human rights.* Paris, France: UN General Assembly.

Veblen, T. (1899/2009). *The theory of the leisure class.* Oxford, UK: Oxford University Press.

A Brief History of Open Educational Resources

T. J. Bliss* and M. Smith

Hewlett Foundation
*tjbliss@hewlett.org

Editors' Commentary

The Hewlett Foundation has been a key supporter of the open movement, donating over US170 million dollars over the past 15 years. In this chapter, authors T. J. Bliss and M. Smith—both of the Hewlett Foundation—ask whether this investment has been worthwhile. To answer this provocative question they trace the history of the open movement itself. They begin in the 1990s with fledging programs that formed the foundation for modern open education. From there, they cover the period they refer to as 'open's adolescence' from 2004 to 2010. Finally, they discuss recent trends in open, Hewlett Foundation funding priorities, and their hopes for the future of the movement.

Introduction

The Open Educational Resources (OER) movement is fifteen years old. This essay reviews OER's history, extraordinary growth, and place in education from the perspective of one current and one former employee of the William and Flora Hewlett Foundation.[1,2] Since 2001, the Hewlett Foundation has provided just over US$170 million to develop and extend the reach and effectiveness of OER. We tell the story of OER's development, provide examples and discuss uses of OER, and sketch its potential as a powerful tool for reducing

---

How to cite this book chapter:
Bliss, T J and Smith, M. 2017. A Brief History of Open Educational Resources. In: Jhangiani, R S and Biswas-Diener, R. (eds.) *Open: The Philosophy and Practices that are Revolutionizing Education and Science.* Pp. 9–27. London: Ubiquity Press. DOI: https://doi.org/10.5334/bbc.b. License: CC-BY 4.0

inequalities of educational opportunity and promoting innovative strategies to improve educational problems. We realize that our viewpoint shapes our discussion and our examples, thus we have deliberately referenced a large number of OER publications from a wide range of authors so the reader may explore materials that may have different perspectives.

The Early History: 1994–2004

Spurred by a 1994 National Science Foundation grant led by James Spohrer, in 1997, the California State University created MERLOT to identify and provide access to mostly free, online curriculum materials for higher education.[3] Soon after, in 1998, David Wiley, an assistant professor at Utah State University, proposed a license for free and open content as an alternative to full copyright.

MERLOT, now with over 40,000 curated and rated items including over 700 separate psychology materials, provided the early means for college teachers to share intellectual content focused on teaching and learning and Wiley's insight made it easy to turn web-based or other educational materials open for others to use.

Simultaneously, Open Access (sharing research and other intellectual content) was growing from a tiny beginning in 1993 to the publication of PLOS in 2001, currently the largest of over 11,000 open journals including upwards of 700 focused on mathematics, and the Budapest Open Access Initiative[4] in 2002, which helped establish open access as a world-wide approach to sharing research. These three extraordinary events set the stage for the rise of OER.

Ironically, however, also in the first two years of the new millennia, many American universities were attempting to sell their academic content, including elite institutions such as Yale, Columbia, and Stanford. Some institutions went the for-profit route while others chose to have their effort not-for-profit.[5] With a few exceptions, all of the major institutions ended their effort within a few years.

But two major universities headed down a different path. At Rice University, engineering professor Richard Baraniuk, frustrated by the inability of the traditional publishing model to produce timely and relevant textbooks, was building Connexions, a web-based platform to facilitate the development and sharing of open source educational content by university professors all over the world. Connexions, which changed its name to OpenStax, now has over 20 free college level textbooks, including psychology, written by authors around the USA and has been projected to have saved students nearly US$40 million.

And at the Massachusetts Institute of Technology (MIT), after a year-long wide-ranging debate, the faculty and administration committed themselves to freely share with the world the content of all of their courses. The idea for MIT OpenCourseWare (MIT OCW) grew out of discussions of the MIT Council on Education Technology in 1999, which was charged with determining how MIT should position itself in the distance learning[6]/e-learning environment, provide a new model for the dissemination of knowledge and collaboration among scholars around the world, and contribute to the 'shared intellectual commons' in academia, which fosters collaboration across MIT and among other scholars.[7] Its resolution was to open course materials so that anyone anywhere could benefit from MIT's knowledge. They audaciously proposed that course materials from all of their courses would be open to students and professors throughout the world. In the MIT OCW catalog today, there are over thirty psychology courses ranging from Introduction to Psychology to Neuroscience and Behavior and The Art and Science of Happiness.

Early in 2001, then-president Charles Vest visited the Mellon and Hewlett foundations requesting support to make as much content from MIT's roughly two thousand courses available freely online. Both foundations quickly provided multimillion dollar grants and the first fifty OpenCourseWare (OCW) courses, ably developed under the leadership of Executive Director Anne Margulies, were online by September 2002.[8] On the west coast at Stanford in 2001, to support the legality of distributing and altering open materials, especially cultural works such as photos and music, Lawrence Lessig, Hal Abelson[9], and Eric Eldred founded Creative Commons, an organization that develops and releases licenses for free and open materials across a wide range of areas. The work of Lessig et al. built on Wiley's earlier efforts.[10]

The Hewlett Foundation originally conceived of the MIT OCW grant as an important but one-time investment. As we considered alternative educational technology investments, however, MIT's powerful moral and ethical stance became more compelling, and by late 2002, we were focusing most of our technology work on providing open content and making it freely available. Thomas Jefferson captured the spirit of what we wanted to accomplish in a letter he wrote in 1813: 'He who receives an idea from me, receives instruction himself without lessening mine; as he who lights his taper at mine, receives light without darkening me.'[11]

The Hewlett Board of Directors never challenged this direction and for almost 15 years has fully supported work on what became OER, even when some of our grants failed. When we first proposed the large MIT grant, Walter Hewlett, then chair of the board, expressed his understanding of MIT's instincts and shared with board members a story about how he had decided to make his music collection free after considering whether to sell it.

Early on, the Hewlett Foundation followed a simple strategy for stimulating open content. We provided grants to major universities beyond MIT, including Harvard, Carnegie-Mellon, Open University UK (OU UK), and Rice, where we gave Baraniuk a grant to help continue his work. We also funded the development of the OER Commons at the Institute for the Study of Knowledge Management in Education (ISKME). We took this direction in part to make a clear statement against the cliché, often cited by critics, that 'you get what you pay for.' Once certain content was made open to the public—MIT course materials, Harvard's creative library collections, and early versions of Carnegie-Mellon's adaptive cognitive tutor courses—that criticism was stifled.[12]

During the same period, we provided Creative Commons with a general support grant and funded both the Organization of Economic Cooperation and Development (OECD) and the United Nations Educational, Scientific and Cultural Organization (UNESCO) to stimulate interest in openly licensed educational materials in the developed and developing world. At a July 2002 UNESCO meeting of developing world nations in Paris, the name Open Educational Resources (OER) was coined and adopted for this new education innovation. While the thirst for openly licensed content was a clear outcome of this meeting, there was also a resounding uniform voice among the developing nations to be joint contributors to the open repository and not just consumers.[13]

What Are OER?

The Hewlett Foundation defines OER as 'teaching, learning, and research resources that reside in the public domain or have been released under an intellectual property license that permits their free use and re-purposing by others.' Creative Commons provides the licensing tools for permitting this free use and re-purposing; Hewlett considers the Creative Commons Attribution (CC BY) license to be the license of choice, allowing for maximal reuse and repurposing of copyrightable educational resources while still acknowledging the creative work of the developer.[14]

David Wiley elaborated on the idea of the permissions granted to an educational resource by an open license:

> 'The term "open content" describes any copyrightable work (traditionally excluding software, which is described by other terms like "open source") that is licensed in a manner that provides users with free and perpetual permission to engage in the 5R activities:
>
> 1. Retain — the right to make, own, and control copies of the content (e.g., download, duplicate, store, and manage)
> 2. Reuse — the right to use the content in a wide range of ways (e.g., in a class, in a study group, on a website, in a video)
> 3. Revise — the right to adapt, adjust, modify, or alter the content itself (e.g., translate the content into another language)
> 4. Remix — the right to combine the original or revised content with other open content to create something new (e.g., incorporate the content into a mashup)
> 5. Redistribute — the right to share copies of the original content, your revisions, or your remixes with others (e.g., give a copy of the content to a friend).'[15]

From Hewlett's perspective, a long-term goal is for an OER to be openly licensed (under a Creative Commons attribution license that includes the 5R activities), as well as technologically accessible and editable using generally available tools, and designed with diverse learners in mind. Deviation from any of these characteristics reduces the relative 'openness' of an educational resource.[16]

Not surprisingly however, much of the material now called OER (including some content funded by the Hewlett Foundation) do not meet all of these criteria. The Creative Commons licenses provide room for the author of the educational resource to restrict certain areas of use—for example, CC BY ND is a license that requires all users to acknowledge the author (BY) and not create derivatives (ND). Because it does not allow alteration of content, CC BY ND reduces much of the usefulness of the resource.

At its most fundamental level an OER has two powerful components: it is available for free to all and it is adaptable to serve the needs of the user. An ND clause on the license removes the second component—no longer may an OER be translated, altered, or mixed with other materials to improve its usefulness for new users. It may not be changed! MIT's OCW does not have ND on their license – faculty and students and others all over the world may adapt it for their own use.

Another somewhat controversial form of license, which is used by MIT OCW, is CC BY NC where the NC restricts commercial use. The NC condition seems on first glance to provide a simple protection against the transformation of open and free to closed and costly. In many instances this is a valid reaction. But one of the challenges for OER developers, including those that take developed OER and adapt it for a new population or purpose, is that it is very difficult to have a sustainable model for development and continuous improvement if there is no way to create a steady stream of funding. Foundations typically fund new innovations like OER for a while but then change their priorities to focus on something else. Even highly endowed universities do not favor activities that cannot support themselves or are not externally supported. The absence of a NC license allows everyone to have the original work for free and to adapt it however they wish and to market it for remuneration if they wish. For a special photograph or painting or a musical piece perhaps a NC license is particularly appropriate. For an open lecture or other piece of educational content, perhaps it is not.

Thus the big tent of licensed educational materials now generally called OER covers many configurations. While Hewlett's ultimate goal is to stimulate high-quality educational content without restrictions other than acknowledgement of the original content developer, we recognize that many developers have difficulties with losing control over their original work. Some may dislike allowing the original work to be modified or balk at offering their materials, which were intended to be free, without restricting users from making money from the new product. While such instincts are natural, the imposed restrictions have several major costs. They hamper people from tapping into their creative nature to directly build on existing materials and constrain access (e.g., by not allowing translations of the original material into other languages). They also limit the possibilities for business models that might sustain and improve the effectiveness of the content. The trade-offs are a struggle for everyone working in the green fields of OER, so while we prefer fully open resources, we understand and welcome the existing diversity.

OER's adolescence: 2004–2010

In the early years of OER before 2004, the Hewlett strategy towards how to lever-age technology to support educational improvement changed from year to year. In 2004–2005 we adopted a more structured approach.[17] At that time, Hewlett's overall goals focused on the promotion of free, useful educational materials for all. We saw this as a long-term effort and structured our work into three parts:

a. *Supporting high-quality OER content providers in the developed and developing worlds.* We deliberately supported many types of OER from different nations and parts of the world, such as OCW, full courses, teacher training, textbooks, lessons, and simulations.[18] During this time period, OER Africa, a project of the South African Institute for Distance Education, was launched under the leadership of Catherine Ngugi in Narobi to support local OER communities across the African continent.

b. *Building infrastructure and removing barriers to OER.* While pipes and standards are important, infrastructure does not refer only to technical supports. We divided the infrastructure concept into three parts—technical, legal/social/cultural, and research—toward the goal of helping to design and motivate a self-sustaining environment that supported the widespread development and use of OER.[19] More precisely, our intent was to advance opportunities for underserved people throughout the world and to stimulate new opportunities for learning and teaching by using the opportunity to adapt and combine OER materials to meet the specific needs of different teachers and students. The range of open materials including powerful simulations, partial lectures, and new assessment tools provided instructors all over the world with powerful tools that may have only been available in premier universities. On the technical side, our grants took into account massive changes in the delivery of information (e.g., increasing access to the internet and the rising importance of handheld devices) and the literature describing how openness can change behavior and expectations (e.g., the Long Tail, Wikipedia, MoveOn). We paid attention to the legal/social/cultural side through support of organizations, including Creative Commons, the internet Archive that regularly captures the entire web and sponsors creative collections of content; Connexions and other platforms to support OER; the Institute for the Study of Knowledge Management in Education (ISKME) to provide an open and easily accessible library of OER;[20] and China Open Resources for Education (CORE) and Lucifer Chu's OOPS organizations to translate materials into Chinese.[21] To improve and better understand the OER movement, we supported the Organisation for Economic C-operation and Development (OECD) and OU UK and their OER research efforts.[22]

c. *Developing a world movement for open education.* To fulfill this strategic goal, we amassed institutional supporters including UNESCO, OECD,

Commonwealth of Learning (COL), and the Asia-Pacific Economic Cooperation;[23] created networks of producers and users of OER; and supported yearly meetings for people engaged in developing and using OER, including the annual Open Education Conference, the Open Education Global Conference, and Hewlett's own yearly OER meeting. We encouraged national governments to open their materials and research by working with leaders in countries and international agencies such as China, India, and the European Union. We also published articles in magazines like *Change*[24] and *Science*[25] and encouraged advocates for government support of OER in nations all around the world.[26]

Entering this phase, much of our attention focused on OER's usefulness at providing knowledge in its original form to those who otherwise might not have access. The implicit goal was to equalize access to disadvantaged and advantaged peoples of the world – in MIT's language to create 'a shared intellectual common.' Our view captured this focus and extended it to K-12 (kindergarten through secondary) schooling and out-of-school learners as well as to higher education.

The open materials were quite diverse. For some free resources, such as the Open Learning Initiative at Carnegie Mellon,[27] the content of the OER was designed to be practically impossible for a user to alter; thus, it was openly licensed but not technologically open. Other content, such as MIT OCW, could be used for educational purposes in any manner as PDF 'snapshots' of existing courses in whole or part. Over time, many MIT professors and others added video, simulations, pictures, and other materials to their websites, which could be used in the original form, altered, or adapted; thus, for example, professors around the world were able to draw on the MIT OCW when they design and teach their courses. The power and the inherent connection between open and adaptation gradually became evident to us during these early years and this knowledge began to influence our selection of grants.

An important benchmark occurred in late 2006. At the request of its board, the Hewlett Foundation supported an extensive review of the OER program conducted by three prominent education and technology experts, Daniel E Atkins, John Seely Brown, and Allen Hammond.[28] The report, published in 2007, looked backward and forward. The authors dug deeply into the OER grants and their products and ultimately were enthusiastic about the progress the Foundation had made and recommended that 'the Hewlett Foundation continue to nurture global open educational resources, but to do so on a larger and more diverse scale and in the context of an even bolder goal—to shape a new culture of learning that is now possible in the digital world.'

This report was presented to the Hewlett Board and gave the Foundation program the legitimacy and impetus to follow its initial strategic plan until 2011. It did so with a special focus on infrastructure, which provided a support for new OER to be created and released with Creative Commons' licenses all around the world with no financial assistance from Hewlett. Other organizations, including the Shuttleworth and Gates Foundations and the Open Society Foundation, had also entered the picture. The Gates Foundation was a supporter of the Khan Academy, perhaps the best known producer of OER other than MIT. Established in 2006 and seriously underway by 2009, Khan's materials have helped hundreds of millions of people learn online, have been translated into 65 languages, and are used in schools and community colleges around the world for remedial and blended learning.[29] Hewlett has not provided support for Khan Academy.

During this period, the world movement in support of OER flourished. In particular, two significant international meetings affirmed OER: the Shuttleworth Foundation supported a meeting in South Africa in 2008 and UNESCO hosted an OER World Congress in Paris in 2012. Each meeting involved representatives from dozens of nations who voiced their commitment to OER. Importantly from our perspective, neither of these meetings was funded or led by the Hewlett Foundation.[30]

Open Access (OA) had also grown rapidly along with OER. The OA initiative responded to the pace of science and the need to ease the path of new knowledge by openly distributing research studies and data. The Scholarly Publishing and Academic Resources Coalition (SPARC), a primary advocate for OA in the United States led by Executive Director Heather Joseph for over a decade, has been a strong partner in the open theater more generally. Specifically, SPARC recently hired a director of Open Education to further its efforts in the OER space. Today, such advocates of OER are found in dozens of nations.[31]

Finally, the early years of this decade saw the rise of Massive Open Online Courses (MOOCs) produced by some of the most well-known universities in the United States and around the world. While free, online courses have existed for years—like the Virtual University of Pakistan, which digitizes and freely disseminates all its classes on YouTube and through other means—MOOCs have captured the imagination of mainstream media. Although 'open' is in the name, only a relatively few MOOCs are free and only a handful carry a Creative Commons license that would allow user institutions or individuals to alter and adapt the content.[32] But even though most MOOCs are not OER, their rise has generated interest in valuable content that has hitherto been impossible for almost all of the world's population to access. In this regard, the MOOCs arguably have contributed positively to the open movement and fall within Jefferson's vision of sharing ideas—certainly the free MOOCs would meet his standard.[33]

A Change in Strategy: 2011–2015

With the release of a new strategic plan in late 2010, Hewlett sought to deepen the movement by 'going mainstream' and focusing more attention on improving educational practice in the United States.[34] Hewlett continued its support of key

OER infrastructure efforts but also provided grants for creating more polished, market-ready primary resource products, such as full end-to-end K–12 curricula and complete textbooks aligned to higher education courses with problem sets and teacher supports. In the United States, as part of its new strategy, Hewlett linked its OER efforts with its priority supporting the implementation of the Common Core State Standards by helping EngageNY, which provides openly licensed Common Core-aligned curricula in K–12 school mathematics and English language arts. To date, the EngageNY curricula have been downloaded more than 20 million times in schools throughout the country and the world.[35]

Hewlett has also promoted the use of open textbooks at the K–12 and college levels to great success. Many of the open textbooks produced by CK–12[36] and K–12 OER Collaborative are used in K–12 schools around the world. Hewlett, Gates, and the California government also support the integration of OER into the existing public higher education system through an organization called the California Open Educational Resources Council, which promotes the use of open textbooks and other materials.[37] At the collegiate level, the use of open textbooks produced by such organizations as OpenStax College, BCcampus, and Lumen Learning has become quite popular, particularly as it helps to reduce students' financial burden.[38]

The mainstream strategy also focused on involving federal governments in the OER movement. Creative Commons and SPARC have led the effort to encourage the government to support open research and the development of open educational products. In fall 2015 the US Department of Education and the White House Office of Science and Technology Policy announced new open policies supportive of the development of OER.[39] At the end of that year, the US Department of Labor announced a regulation requiring all intellectual property developed under a competitive Labor Department grant be released with a CC BY license.[40]

The United States is not alone—governmental adoption of OER is moving quickly throughout the developed world. In a book released in 2015, the OECD reports: 'In August and September 2014, governments were asked to respond to a CERI/OECD questionnaire on how they support and facilitate the development and use of OER in all education sectors. The survey collected the responses of 33 countries: 29 OECD member countries and 4 accession and key partner countries (Brazil, China, Indonesia and Latvia). The results indicate a clear policy support for OER, with 25 countries reporting having a government policy to support OER production and use.'[41]

Has It Been Worth US$170 Million?

Hewlett's commitment to OER is not over, but it has been 15 years since we first funded MIT OCW—a substantial amount of time—so we should consider whether our investment in OER has been worth it.

This is an existential question—the answer depends on who we are asking and on what an imagined alternative might have produced. One consideration is that just over US$170 million already spent developing OER has led to positive outcomes and will also produce future benefits. Another consideration has to do with values and goals of the Foundation, which emphasize the 'well-being of mankind' and the support of practical innovation. A third consideration is that although this effort was heralded by some as a 'magic bullet' that would easily solve complex problems in education, this proved not to be the case.

Hewlett has never treated its support for OER as a short-term trial, one to be dropped as priorities changed. Rather, throughout the years, Hewlett has remained committed to its original goals of making OER a powerful tool to improve the equality and quality of educational opportunity around the world. Since 2003 the Foundation has treated its grants in OER as providing support for a social movement that in time should be self-sustainable.[42] The steady increase of nations adopting legislation or regulation that support and sometimes require the use of open licenses is one measure of positive growth of self-sustainability. While the OER movement has not fully achieved scale, it is well on its way.

The extent of coverage for higher education from MIT OCW and its many translations into different languages is enormous. More than 100 million unique visitors, including scholars, teachers, and students, have explored content on the MIT site (and millions more who speak and read in languages other than English have visited the sites of the 250 higher education institutions from all over the world in the Open Education Consortium). Previously, only students who could afford four years at MIT or another elite institution would have been able to access the OCW content, but now professors, students, and people all over the world can draw on these resources for knowledge. Thousands of open textbooks and hundreds of full open courses are now available for the most highly enrolled US college courses and are being translated into many languages,[43] helping more students afford college.[44] PhET science simulations have been downloaded over 275 million times. Teacher Education in Sub-Sahara Africa (TESSA), COL, and TESS-India support high-quality professional development for teachers in a half dozen African nations and seven states in India, together influencing teachers of hundreds of thousands of students.[45]

At the K–12 level, Khan Academy materials have had hundreds of millions of users. An important byproduct is the work of the Foundation for Learning Equality (FLE); FLE has developed a method for delivering Khan and other educational materials in settings where there is no connectivity and no electrical power.[46] FLE has brought educational materials to an estimated 2 million-plus users through its work with large non-governmental organisations (NGOs) in refugee camps, US prisons, and other resource-limited settings. Full K–12 open curricula reaching millions of students are available in English, as are hundreds of textbooks and online courses. Open textbooks are also available in dozens of countries.[47]

There are still barriers and problems in the OER world even though many nations have endorsed the use of OER. Surveys indicate that only a small percentage of professors and teachers know very much about OER even if they use open materials. We expect this number to improve over time, however, as teachers and professors adapt and draw on the openly licensed materials. Another potential barrier comes from the publishing industry because, as a practice, the use of OER threatens their business model. Still, we expect that the industry's modes of delivery and possible sources of income will adapt to OER. This trend is pushed by models for curricula changing more toward using the internet, examples of MOOCs and OCW becoming more evident and teachers developing a greater understanding of ways to adapt materials to their students. It is clear that the age of the bound textbook that stays unchanged for six years will soon be over.

These and other problems are real challenges but they now seem solvable. This optimism and the strong increase in the raw numbers for the creation and use of OER as well as the positive activity at the government level around the world indicate a healthy, useful and vibrant OER movement.

The Next Stage: OER Helping to Solve Problems: 2016 and Beyond

We believe that the OER movement now has staying power without major support from the Hewlett Foundation. One of many signs of independence of the movement is that a group of OER activists and leaders recently published a living OER strategy document that may be adapted and modified and is designed to address 'strategic questions about how we, as the global OER movement, can reach our collective goals.'[48]

In late 2015 the Foundation released a refreshed strategic approach for its OER portfolio, describing Hewlett's three goals: strengthening infrastructure, using OER to help solve social and educational problems, and improving educational materials.

The focus on infrastructure will include supporting institutions such as Creative Commons and ISKME that have been mainstays of the OER movement and on increasing the quality and quantity of research projects and descriptions of OERs use and effectiveness. The OU UK, OECD, and UNESCO already provide a steady stream of useful description and research on OER, while the International Development Research Centre in Canada supports local researchers in the developing world who study the use and effectiveness of OER interventions.[49] It is important to note that work carried out by local researchers might be especially useful in the developing world because the findings would have regional credibility. Such infrastructure involvement also has the side benefit of supporting Open Access, Open Culture, and Open Government.

For the second goal Hewlett will fund tailored, innovative interventions and strategies that use OER as a tool to help solve social and educational problems.

On top of the list are traditional problems of inequity and opportunity. Since these problems vary, the contribution of OER also may vary. OER cannot address or ameliorate all of the inequalities in the access to knowledge and education but with careful and sustained attention, some may be lessened.

At the heart of the concept of OER is freedom: freedom of access to content, freedom from cost, and freedom to use in any way. Large classes of people in the United States and across the globe do not receive an adequate education due to a lack of finances or other resources. By providing free access to powerful education content, OER can help underserved populations such as children and youth in prison or foster care, Native Americans in government schools, and students from low-income families or who must learn English as a second language. When it comes to tertiary education, which has a high cost particularly in the United States, the financial barrier may be partially offset by high-quality open textbooks and online courses. Though Hewlett has already supported OER work in some of these settings, much more can be done.

Even more egregious problems exist overseas, particularly in the developing world. The Foundation has been allocating resources into this area, for instance, supporting FLE in its work with large intergovernmental organisations (IGOs) to provide education in areas that lack connectivity and sometimes even electrical power. These settings, which include refugee camps and thousands of tiny, low-priced, low-cost private schools in East Africa, Pakistan, and India, are in dire need of open materials and effective delivery systems.[50] Another problem is teacher training, which is often widely neglected in many parts of the developing world; TESSA in Africa and TESS-India provide existing OER models that also could be used in many other settings. And the OER Commons, developed by ISKME founder and CEO Lisa Petrides, has expanded its focus to extend to educators across the globe.

Such successes suggest other strategies that could work in locations in the developing and developed world. Open MOOCs and video could be used for pre-service training for teachers. Many institutions, especially in low-income countries, lack laboratories for science experiments and medical diagnostic opportunities. This deficiency might be partially offset by providing these institutions access to open, high-quality virtual laboratories, diagnostic rooms, and operating rooms. Another more ambitious example supported by Hewlett might be the Peer to Peer University (P2PU), an open, free institution that provides free courses and largely relies on the power of sharing and collaboration among its students. Now may be the time to consider whether this innovation is working and, if it is working well, to expand or replicate it to meet global needs. All of these efforts exemplify the kind of OER-focused work that will help balance the equality scales.

Hewlett's third goal in the OER arena is to improve the quality and usefulness of educational materials. As free platforms making it easier to adapt and mix content become more usable, teachers and other educators can piece together OER from multiple sources to create curriculum geared toward the needs of their specific classrooms and schools. Professional networks of teachers and others who openly share extend this power of the freedom to collaborate and build. The growing capacity and ease of translating materials from one language into others further increases the scope of materials available at the local level. Such efforts all support the ability of OER to bring about continuous improvement and innovation.

A related, potentially powerful and innovative approach to improving content quality was sketched out in the 2007 independent review of Hewlett's OER program. Thinking on a big scale, the two authors argued: 'We believe that the Hewlett Foundation can play a leadership role in weaving the threads of an expanded OER movement; the e-science movement; the e-humanities movement; new forms of participation around Web 2.0; social software; virtualization; and multimode, multimedia documents into a transformative open participatory learning infrastructure—the platform for a culture of learning.'

Atkins et al. (2007) sketched the dimensions of their vision of an infrastructure of learning built around OER.[51] Perhaps this is only an idealized vision but in 2016 it may well be a vision worth exploring. It fits the quality, usefulness, and big-problems criteria of the OER agenda.[52]

A Final Word

The world of OER is vibrant, challenging, and filled with tremendous possibilities. To quote the new vision document: 'The Hewlett Foundation is excited to continue supporting OER at a time that the field is building on its successes and transitions to solving some of the most pressing problems that teachers and students face throughout the world. With this new problem-based approach, the Foundation looks forward to many more students benefitting from the promise of OER.'

Notes

[1] Other Hewlett employees directly responsible for elements of the OER portfolio referenced throughout this paper include Catherine M. Casserly, program officer/OER initiative director (2001–2009); Kathy N. Grant, associate program officer/program officer (2009–2014); Vic Vuchic, program officer (2007–2014); and, Dana Schmidt, OER program officer (2014–2016).

[2] We thank Catherine Casserly and Victor Vuchic for their comments on earlier versions of this text. We also thank Vijay Kumar of MIT and Cable Green of Creative Commons, both important OER advocates, for their comments.

[3] The grant was 'Authoring Tools and an Educational Object Economy' (EOE) and was led by James Spohrer and hosted by Apple Computer and other industry, university, and government collaborators. The EOE developed and

distributed tools to enable the formation of communities engaged in building shared knowledge bases of learning materials. MERLOT, the acronym for Multimedia Educational Resources for Learning and Online Teaching, is a robust site with links to over 40,000 teaching and learning resources. MERLOT maintains a link for each resource as well as metadata, which contain information about the cost (if any) and the use permissions for each of the materials. Simultaneous a federal working group led by Kurt Winters in the Department of Education in response to a memo from President Bill Clinton conceived the FREE initiative in 1997 which collected and made available the most highly rated free government educational content, see http://clinton2.nara.gov/WH/New/NetDay/memorandum.html. The FREE site was closed in June 2015.

4   For more information, please see http://www.soros.org/openaccess/read. shtml.

5   'Oxford, Yale and Stanford closed their joint not-for-profit online venture, AllLearn (Alliance for Lifelong Learning) citing insufficient enrollments and funding as the primary reasons. AllLearn was established in 2001 at the peak of the dot-com boom to offer online non-credit courses in general interest subject areas. The initial audience was the alumni of the three institutions, but as of the autumn semester 2002, provision was opened to the general public. After almost five years in operation, the three universities have released a joint statement concluding that 'the cost of offering top-quality enrichment courses at affordable prices was not sustainable over time.' Following a series of collapsed e-university ventures from US universities (e.g., NYU Online, Fathom, Virtual Temple, and University of Maryland University College Online), AllLearn is another major product of the dot-com boom to fold.' See https://ox-d.promediagrp.com/w/1.0/afr?mi=ec00476b-a355-4ab0-fa02-ba6a24422bb9&ma=1454266450&mr=1455476050&mn=1&mc=1&cc=1&auid=33998&cb=0.36133100.1454266136.

6   For more information, please see https://en.wikipedia.org/wiki/Distance_ learning.

7   See https://en.wikipedia.org/wiki/MIT_OpenCourseWare.

8   See https://cnx.org and http://openstax.org for the Connexions project (now called OpenStax) at Rice, and see www.creativecommons.org for Creative Commons. See http://ocw.mit.edu/index.htm for MIT OCW.

9   For more information, please see https://en.wikipedia.org/wiki/ Hal_Abelson.

10   For more about the Creative Commons, see http://creativecommons.org.

11   Jefferson's letter went on: 'That ideas should freely spread from one to another over the globe, for the moral and mutual instruction of man, and improvement of his condition, seems to have been peculiarly and benevolently designed by nature, when she made them, like fire, expansible over all space, without lessening their density in any point, and like the air in which we breathe, move, and have our physical being, incapable of confinement

or exclusive appropriation. Inventions then cannot, in nature, be a subject of property.' See http://press-pubs.uchicago.edu/founders/documents/a1_8_8s12.html for the full text.

[12] For the Harvard collections, see http://ocp.hul.harvard.edu; for the Carnegie-Mellon Online Course Initiative, see http://oli.cmu.edu; and for the Connexion project at Rice, see cnx.org.

[13] See the UNESCO meeting report: http://www.hewlett.org/uploads/files/OpenCourseWareandDevelopingCountries.pdf. See also OECD at http://www.oecd.org/edu/ceri/givingknowledgeforfreetheemergenceofopeneducationalresources.htm and UNESCO work on OER at http://www.unesco.org/new/en/communication-and-information/access-to-knowledge/open-educational-resources/.

[14] While the CC BY license is often thought of as the least restrictive Creative Commons license, it requires acknowledgement of the originator. When there are multiple serial redesigns of the same content, even the CC BY license becomes awkward; thus there is an even less restrictive license (CC0), which eliminates the acknowledgement requirement.

[15] Retrieved from http://opencontent.org/definition/ on December 4, 2015.

[16] Other aspects of openness beyond legality are often considered with OER, including technological openness and accessibility. Technological openness refers to the freedom available to an end user to engage in the 5R activities. For example, content that is legally open but hidden behind a paywall or distributed in a format that restricts repurposing is not within the spirit of open content. Similarly, content that is openly licensed and technologically open but not designed with general accessibility in mind is not fully open content. Jutta Treviranus, the director of the Inclusive Design Research Centre at the Ontario College of Art and Design University, explains, 'Correctly designed digital resources can transform to the unique specifications of each learner, presenting the visual layout, presentation modes (e.g., audio, visual, tactile), and method of control that suits the individual learner.' Treviranus argues further that OER should be designed for diverse learners, not just for the typical student. Personal communication to T. J. Bliss, July 2015.

[17] The metrics for OER growth are difficult. The number of hits for the term *Open Educational Resources* was zero in a Google search in 2002, in the tens of thousands in 2006–2007 when we were watching it, and now is in the hundreds of thousands. It is even more difficult to measure the term OER, which now has more than 20 million hits with more than 90 percent of the first ten pages referring to Open Educational Resources. Until the fall of 2005, almost the entire first page of a Google search for Open Educational Resources directed to sites that Hewlett funded—by mid-2006, fewer than half of the citations on the first page referred to Hewlett-supported projects. See http://www.hewlett.org/uploads/files/OER_overview.pdf for an interesting report of the work up through 2005.

[18] For information about OER work from 2004–2010 at various institutions, see: http://www.open.edu/openlearn/ for the Open University of the United Kingdom (OU UK); https://www.col.org/news/speeches-presentations/open-education-resources-oer-what-why-how for the Commonwealth of Learning (COL) work on OER; http://www.oeconsortium.org for the Open Education Consortium; http://www.oecd.org/edu/ceri/38149140.pdf for the Open University of the Netherlands; https://phet.colorado.edu for PhET's interactive science simulations; http://www.tessafrica.net for TESSA; http://iite.unesco.org/pics../publications/en/files/3214700.pdf and https://en.wikipedia.org/wiki/China_Open_Resources_for_Education for CORE; and http://www.oerafrica.org for OER Africa.

[19] Without being aware of it, our approach was similar to Edwards et al.'s [2007] theory of infrastructure, which requires attention to legal, technical, cultural, social, political, and financial components. See Paul N. Edwards, Steven I. Jackson, Geoffrey C. Bowker, and Cory P. Knobel, 'Understanding Infrastructure: Dynamics, Tensions, and Design.' January 2007. http://deepblue.lib.umich.edu/handle/2027.42/49353. See also Marshall S. Smith and Phoenix Wang, 'The Infrastructure of Open Educational Resources.' *Educational Technology*, v. 47, n. 6, pp. 10–14, Nov.–Dec. 2007.

[20] ISKME is the Institute for the Study of Knowledge Management in Education. See http://www.iskme.org.

[21] OOPS is a volunteer-based localization project with the goal of translating open knowledge into Chinese. More than 20,000 volunteers are estimated to have joined OOPS (Opensource Opencourseware Prototype System), which translates MIT and other OCW into Mandarin. CORE is China Open Resources in Education. For a longer discussion of OER and China, see http://iite.unesco.org/pics../publications/en/files/3214700.pdf and for more information about translations of MIT OCW and other OCW, see http://ocw.mit.edu/courses/translated-courses/.

[22] See, for example, http://www.oecd.org/edu/ceri/38654317.pdf and http://www.open.ac.uk/score/events/learning-oer-research-projects. For a more recent discussion of research, see http://www.hewlett.org/research-open-oer-research-hub-review-futures-research-oer and http://www.hewlett.org/sites/default/files/OER%20Research%20paper%20December%2015%202013%20Marshall%20Smith_1.pdf.

[23] See, for example, https://www.col.org/, http://www.nba.co.za/asia-pacific-economic-cooperation-apec-oer and http://www.apec.org.

[24] See Marshall S. Smith and Catherine M. Casserly, 'The Promise of Open Educational Resources.' *Change: The Magazine of Higher Learning*, v. 38, n. 5 pp. 8–17, Sept.–Oct. 2006.

[25] See Marshall S. Smith, 'Opening Education.' *Science* v. 323, n. 89, pp. 89–93, Jan. 2, 2009.

[26] For an example of advocacy, Hewlett funding helped support Vijay Kumar's work with India's National Knowledge Commission, leading to

strategic recommendations to use OER to extend access to quality educational opportunity in India. See http://eprints.rclis.org/7462/1/National_ Knowledge_Commission_Overview.pdf and http://www.sampitroda.com/ knowledge-commission/. Also see http://iite.unesco.org/pics../publications/ en/files/3214700.pdf for information about the adoption of OER in various nations.

27 See http://oli.cmu.edu.

28 See *A Review of the Open Educational Resources (OER) Movement: Achievement, Challenges, and New Opportunities* (2007) at http://www.hewlett. org/uploads/files/ReviewoftheOERMovement.pdf, authored by Daniel E. Atkins, professor of Information, Computer Science, and Electrical Engineering at the University of Michigan and former director of the Office of Cyberinfrastructure, US National Science Foundation; John Seely Brown, former chief scientist of Xerox and director of its Palo Alto Research Center; and Allen Hammond, former vice president for Innovation and Special Projects at World Resources Institute.

29 See khanacademy.org.

30 For a discussion of government policies and OER, see http://www.slide share.net/oeconsortium/impact-of-international-organizations-on-governmental-oer-policies-48013353. For more information on the meetings, see https://en.wikipedia.org/wiki/Cape_Town_Open_Education_Declaration and the resolution from the 2012 UNESCO World OER Congress: http:// ru.iite.unesco.org/files/news/639202/Paris%20OER%20Declaration_ 01.pdf.

31 See http://www.sparc.arl.org and https://en.wikipedia.org/wiki/Open_access.

32 As always, there are a few exceptions. For example, the online learning destination edX makes many of its MOOCs free but not alterable while the edX platform itself has open-source code, which is free to use. In 2015 Creative Commons worked with edX to make it simple to add a CC license to edX MOOCs.

33 For more about MOOCs, see https://en.wikipedia.org/wiki/Massive_open_ online_course.

34 See Hewlett Education Program Plan Fall 2010: http://www.hewlett.org/ uploads/documents/Education_Strategic_Plan_2010.pdf. To support the change, the Foundation noted, 'After eight years of field building with the Hewlett Foundation as its primary supporter, OER is beginning to shift from a nascent movement to a respected force in education. The movement was featured in both the *New York Times* and *Wired* magazine in 2010. At the same time, other foundations dramatically increased OER funding, and federal grant programs began to include OER as a priority in their grant application guidelines, signaling greater acceptance of the field.'

35 See http://www.edweek.org/ew/articles/2015/06/10/ny-open-education-effort-draws-users-nationwide.html.

36 See http://www.ck12.org/student/.

[37] See http://icas-ca.org/coerc.

[38] See http://www.ck12.org/student/, http://k12oercollaborative.org/ https://openstaxcollege.org/, http://bccampus.ca/, and http://lumenlearning.com.

[39] See http://www.ed.gov/news/press-releases/us-department-education-launches-campaign-encourage-schools-goopen-educational-resources.

[40] See http://ies.ed.gov/ncee/wwc/quickreview.aspx?sid=245.

[41] See Dominic Orr, Michele Rimini, and Dirk van Damme, *Open Educational Resources: A Catalyst for Innovation*. Educational Research and Innovation, OECD Publishing. Paris, 2015, p. 20. See http://www.oecd-ilibrary.org/docserver/download/9615061e.pdf?expires=1451782268&id=id&accname=guest&checksum=A8C62EF8A123FB6D46F068B779098A8E.

[42] In 2006 Gary Matkin from the University of California at Irvine wrote about OER as a movement. See 'The Open Educational Resources Movement: Current Status and Prospects' at http://unex.uci.edu/pdfs/dean/matkin_apru_paper.pdf.

[43] For examples of open higher education textbooks in psychology, see the following web sites as of Nov. 30, 2015. This is only a sampling of the material on the web: https://open.umn.edu/opentextbooks/SearchResults.aspx?subjectAreaId=7, http://nobaproject.com, http://library.calstate.edu/textbook/?isbn=9780073532073&button=Search&type=books, http://cnx.org and http://ocw.uci.edu.

[44] For example, see references to the importance to students of free textbooks in two major mass media outlets. See http://www.usnews.com/education/best-colleges/paying-for-college/articles/2013/08/14/4-ways-to-get-free-college-textbooks and http://www.cnn.com/2014/04/18/living/open-textbooks-online-education-resources/. For an example of the use of textbooks in other parts of the world, see https://www.col.org/programmes/technology-enabled-learning/eastern-caribbean-open-textbook-forum-supports-oer-strategies.

[45] See http://www.tessafrica.net, https://www.col.org/programmes/teacher-education, and http://www.tess-india.edu.in.

[46] See Foundation for Learning Equality at https://learningequality.org.

[47] For an interesting review of the use of OER in the British Commonwealth, see https://www.col.org/news/speeches-presentations/open-education-resources-oer-what-why-how. For examples of use of open textbooks in the United States, see http://www.ck12.org/; http://www.uen.org/oer/. For an example of the use of open textbooks and professional development in Africa, see http://www.siyavula.com/.

[48] oerstrategy.org, 'Foundations for OER Strategy Development.' Version 1.0. Nov. 18, 2015. Drafting Committee: Nicole Allen, Delia Browne, Mary Lou Forward, Cable Green, and Alex Tarkowski. See http://www.oerstrategy.org/home/read-the-doc/.

[49] See http://www.idrc.ca/en/themes/information_and_communication/pages/projectdetails.aspx?projectnumber=107311.

[50] Global Business Coalition for Education (2016) Exploring the Potential of Technology to Deliver Education & Skills to Syrian Refugee Youth. See http://gbc-education.org/wp-content/uploads/2016/02/Tech_Report_ONLINE.pdf

[51] See Atkins, D., Brown, J. & Hammond, A. (2007). A review of the open educational resources (OER) movement: Achievements, challenges, and new opportunities (pp 1–84). A report to the William and Flora Hewlett Foundation. See endnote 25.

[52] We strongly recommend the CERI/OECD report referenced earlier. It is a tour de force of ideas, projects, and improvements that would enliven the OER movement. We have touched upon many of them independently but the CERI/OECD treatment is more extensive. See http://www.oecd-ilibrary.org/docserver/download/9615061e.pdf?expires=1451782268&id=id&accname=guest&checksum=A8C62EF8A123FB6D46F068B779098A8E.

Open Licensing and Open Education Licensing Policy

Cable Green

Creative Commons, cable.green@gmail.com

Editors' Commentary

It would not be an overstatement to say that Creative Commons licenses provide the legal foundation for most of the open education movement. These licenses— free and easy to apply—provide educators, scholars, and artists the language with which to share their work on their own terms. In this chapter, author Cable Green provides a primer on the licenses themselves before going on to explore how public policymakers can leverage open licensing policies to effectively combat a range of challenges including high textbook costs and publicly-funded-yet-paywalled research.

Introduction

I work at Creative Commons (CC), as the Director of Open Education, because I seek to create a world in which the public has free, legal and unfettered access to effective, high quality education and research resources, and learning opportunities. I've spent my career working in post-secondary education and have seen students: take fewer courses because of the high cost of textbooks, go without required educational resources due to cost, and graduate with tens of thousands in debt. After learning about 'open education,' I decided to join the movement and help more learners access affordable, meaningful learning opportunities.

How to cite this book chapter:
Green, C. 2017. Open Licensing and Open Education Licensing Policy. In: Jhangiani, R S and Biswas-Diener, R. (eds.) *Open: The Philosophy and Practices that are Revolutionizing Education and Science.* Pp. 29–41. London: Ubiquity Press. DOI: https://doi.org/10.5334/bbc.c. License: CC-BY 4.0

Open education is an idea, a set of content and a community which, properly leveraged, can help everyone in the world access free, high quality, open learning materials for the marginal cost of zero. We live in an age of information abundance where everyone, for the first time in human history, can potentially attain all the education they desire. The key to this sea change in learning is Open Educational Resources (OER). OER are educational materials that are distributed at no cost with legal permissions for the public to freely use, share, and build upon the content. The Hewlett Foundation defines OER as teaching, learning, and research resources that reside in the public domain or have been released under an intellectual property license that permits their free use and re-purposing by others.[1] OER are possible because:

- educational resources are digital[2] and digital resources can be stored, copied, and distributed for near zero cost;
- the internet makes it simple for the public to share digital content; and
- Creative Commons licenses (and public domain tools) make it simple and legal to keep one's copyright and legally share educational resources with the world.

Today we can share effective education materials with the world for near zero cost. As such, I argue educators and governments supporting public education have a moral and ethical obligation to do so. After all, education is fundamentally about sharing knowledge and ideas. I believe OER will replace much of the expensive, proprietary content used in academic courses – it's only a matter of time. Shifting to this model will generate more equitable economic opportunities globally and social benefits without sacrificing quality of educational content. In this chapter, I will first discuss how 'open licensing' works and why it is a critical part of OER. We will then explore how and why governments and foundations (funders) are starting to use open educational licensing policies to require open licenses on educational resources they fund.

Open Licensing

Long before the internet was conceived, copyright law regulated the very activities the internet, cheap disc space and cloud computing make essentially free (copying, storing, and distributing). Consequently, the internet was born at a severe disadvantage, as preexisting copyright laws discouraged the public from realizing the full potential of the network.

Since the invention of the internet, copyright law has been 'strengthened' to further restrict the public's legal rights to copy and share on the internet[3]. For example, in 2012 the US Supreme Court on upheld the US Congress's right to extend copyright protection to millions of books, films, and musical compositions by foreign artists that once were free for public use. Lawrence Golan, a

University of Denver music professor and conductor who challenged the law on behalf of fellow conductors, academics and film historians said 'they could no long afford to play such works as Sergei Prokofiev's "Peter and the Wolf," which once was in the public domain but received copyright protection that significantly increased its cost.'[4]

While existing laws, old business models, and education content procurement practices make it difficult for teachers and learners to leverage the full power of the internet to access high-quality, affordable learning materials, OER can be freely retained (keep a copy), reused (use as is), revised (adapt, adjust, modify), remixed (mashup different content to create something new), and redistributed (share copies with others)[5] without breaking copyright law. OER allow the full technical power of the internet to be brought to bear on education. OER allow exactly what the internet enables: free sharing of educational resources with the world.[6]

What makes this legal sharing possible? Open licenses. The importance of open licensing in OER is simple. The key distinguishing characteristic of OER is its intellectual property license and the legal permissions the license grants the public to use, modify, and share it. If an educational resource is not clearly marked as being in the public domain or having an open license, it is not an OER. Some educators think sharing their digital resources online, for free, makes their content OER – it does not. Though it is OER if they go the extra step and add an open license to their work.

The most common way to openly license copyrighted education materials – making them OER – is to add a Creative Commons[7] license to the educational resource. CC licenses are standardized, free-to-use, open copyright licenses that have already been applied to more than 1.2 billion copyrighted works across 9 million websites.[8]

Collectively, CC licensed works constitute a class of educational works that are explicitly meant to be legally shared and reused with few restrictions. David Bollier writes:

> 'Like free software, the CC licenses paradoxically rely upon copyright law to legally protect the commons. The licenses use the rights of owner-ship granted by copyright law not to exclude others, but to invite them to share. The licenses recognize authors' interests in owning and con-trolling their work — but they also recognize that new creativity owes many social and intergenerational debts. Creativity is not something that emanates solely from the mind of the "romantic author," as copyright mythology has it; it also derives from artistic communities and previ-ous generations of authors and artists. The CC licenses provide a legal means to allow works to circulate so that people can create something new. Share, reuse, and remix, legally, as Creative Commons puts it.'[9]

While custom copyright licenses can be developed to facilitate the develop-ment and use of OER, it may be easier to apply free-to-use, global standardized

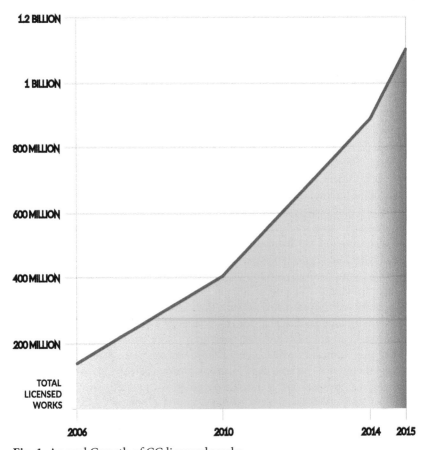

Fig. 1: Annual Growth of CC licensed works.

licenses developed specifically for that purpose, such as those developed by Creative Commons.[10]

Creative Commons Licenses

Because definitions of OER place such an emphasis on copyright permissions and licensing, a basic understanding Creative Commons licenses is critical to understanding OER. CCs open copyright licenses and tools forge a balance – allowing copyright holders to share their work – inside the traditional 'all rights reserved' setting that copyright law creates. CC licenses give everyone from individual creators to large companies and institutions a simple, standardized way to grant copyright permissions to their creative work.

All Creative Commons licenses have many important features in common:

- Every CC license helps creators retain copyright while allowing others to copy, distribute, and make some uses of their work – at least non-commercially.
- Every CC license also ensures licensors get the credit (attribution) for their work.
- Every CC license works around the world and lasts as long as applicable copyright lasts (because they are built on copyright).

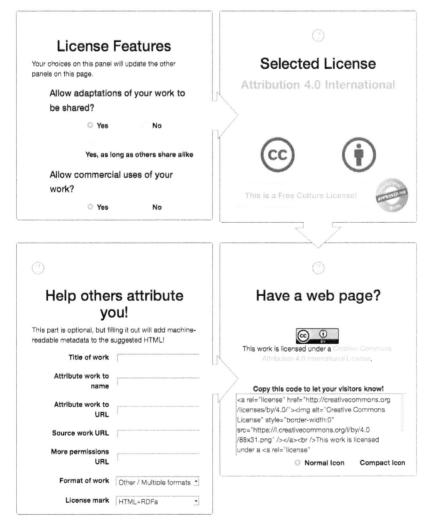

Fig. 2: Registering a CC licensee.

These common features serve as the baseline, on top of which authors can choose to grant additional permissions when deciding how they want their work to be used.

CC licenses do not affect freedoms that the law grants to users of creative works otherwise protected by copyright, such as exceptions and limitations to copyright law like fair dealing or fair use rights. CC licenses require the public to get permission to do any of the things with a work that the law reserves exclusively to a copyright holder and that the license does not expressly allow. Users of a CC licensed work must credit the author; keep copyright notices intact on all copies of the work, and link to the CC license deed (e.g., CC BY 4.0) from copies of the work. Users of CC licensed works also cannot use technological measures to restrict access to the work by others. For example, I cannot lock down your CC licensed music with digital rights management software to restrict others' use.

Anyone can get their CC license – at no cost – at CC's license chooser: http://creativecommons.org/choose It is worth mentioning there is no need to register your work to get a CC license.

The Licenses[11]

Fig. 3: The CC-BY license.

Attribution: CC BY

View License Deed | View Legal Code

This license lets others distribute, remix, tweak, and build upon your work, even commercially, as long as they credit you for the original creation. This is the most accommodating of licenses offered. Recommended for maximum dissemination and use of licensed materials. This is the license required by the US Department of Labor on all of their grants, the Campus Alberta OER initiative,[12] BC Open Textbooks Project,[13] and hundreds of other OER projects around the world. CC BY is recommended for most open licensing policies, and for OER when the author wants to maximize reuse and remix of their work.

Fig. 4: The CC-BY-Share Alike license.

Attribution-ShareAlike: CC BY-SA

View License Deed | View Legal Code

This license lets others remix, tweak, and build upon your work even for commercial purposes, as long as they credit you and license their new creations under the identical terms. This license is often compared to 'copyleft' free and open source software licenses. All new works based on yours will carry the same license, so any derivatives will also allow commercial use. This is the license used by Wikipedia, and is recommended for materials that would benefit from incorporating content from Wikipedia and similarly licensed projects.

Fig. 5: The CC-BY-Non Commercial Use license.

Attribution-NonCommercial: CC BY-NC

View License Deed | View Legal Code

This license lets others remix, tweak, and build upon your work non-commercially, and although their new works must also acknowledge you and be non-commercial, they don't have to license their derivative works on the same terms. Authors use this license when they are fine with free reuse, but not commercial uses of their work.

Fig. 6: The CC-BY-Non Commercial Use-Share Alike license.

Attribution-NonCommercial-ShareAlike: CC BY-NC-SA

View License Deed | View Legal Code

This license lets others remix, tweak, and build upon your work non-commercially, as long as they credit you and license their new creations under the identical terms. MIT's OpenCourseWare project and the Khan Academy both use this license.

Fig. 7: The CC-BY-No Derivative works license.

Attribution-NoDerivs: CC BY-ND

View License Deed | View Legal Code

This license allows for redistribution, commercial and non-commercial, as long as it is passed along unchanged and in whole, with credit to you. This is not an OER compatible open license because the ND clause doesn't allow others to revise or remix the work.

Fig. 8: The CC-BY-Non Commercial Use- No Derivative works license.

Attribution-NonCommercial-NoDerivs: CC BY-NC-ND

View License Deed | View Legal Code

This license is the most restrictive of our six main licenses, only allowing others to download your works and share them with others as long as they credit you, but they cannot change them in any way or use them commercially. This is not an OER compatible open license because the ND clause does not allow others to revise or remix the work.

CC also provides tools that work in the 'all rights granted' space of the public domain. CCs CC0 tool allows licensors to waive all rights and place a work in the public domain, and the Public Domain Mark allows any web user to 'mark' a work as being in the public domain.

For OER, the use of CC licenses looks like this:

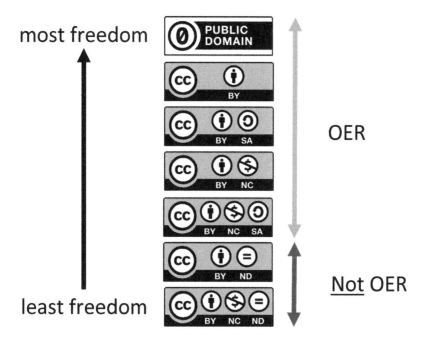

Fig. 9: CC licenses arranged from most to least permissive.

The two CC No Derivatives (ND) are not OER compatible licenses because they do not let the public revise or remix an educational resource. Because the ND licenses violate the 5Rs and every major OER definition, the open education movement does not call ND licensed educational resources 'OER.'

Now that we know what OER is and the role of open licensing in making OER 'open,' the next question is how to make OER the default content produced, adopted, used, and revised in education.

Open Education Licensing Policy

This section explores how public policymakers can leverage open licensing policies, and by extension OER, as a solution to high textbook costs, out-of-date educational resources and disappearing access to expensive, DRM[14] protected e-books. Education policy is about solving education problems for the public. If one of the roles of government is to ensure all of its citizens have access to effective, high-quality educational resources, then governments ought to employ current, proven legal, technical, and policy tools to ensure the most efficient and impactful use of public education funding.

Open education policies are laws, rules, and courses of action that facilitate the creation, use or improvement of OER. While this chapter only deals with open education licensing policies, there has also been significant open education resource-based (allocate resources directly to support OER), inducement (call for or incentivize actions to support OER), and framework (create pathways or remove barriers for action to support OER) open education policy work.[15]

Open education licensing policies insert open licensing requirements into existing funding systems (e.g., grants, contracts, or other agreements) that create educational resources, thereby making the content OER, and shifting the default on publicly funded educational resources from 'closed' to 'open.' This is a particularly strong education policy argument: if the public pays for education resources, the public should have the right to access and use those resources at no additional cost and with the full spectrum of legal rights necessary to engage in 5R activities.

My friend David Wiley likes to say 'if you buy one, you should get one.' David, like most of us, believes that when you buy something, you should actually get the thing you paid for. Provincial/state and national governments frequently fund the development of education and research resources through grants funded with taxpayer dollars. In other words, when a government gives a grant to a university to produce a water security degree program, you and I have already paid for it. Unfortunately, it is almost always the case that these publicly funded educational resources are commercialized in such a way that access is restricted to those who are willing to pay for them a second time. Why should we be required to pay a second time for the thing we've already paid for?[16]

Governments and other funding entities that wish to maximize the impacts of their education investments are moving toward open education licensing policies. National, provincial/state governments, and education systems all play a critical role in setting policies that drive education investments and have an interest in ensuring that public funding of education makes a meaningful, cost-effective contribution to socioeconomic development. Given this role, these policy-making entities are ideally positioned to require recipients of public funding to produce educational resources under an open license.

Let us be specific. Governments, foundations, and education systems/institutions can and should implement open education licensing policies by requiring open licenses on the educational resources produced with their funding. Strong open licensing policies make open licensing mandatory and apply a clear definition for open license, ideally using the Creative Commons Attribution (CC BY) license that grants full reuse rights provided the original author is attributed.

The good news is open education policies are happening! In June 2012, UNESCO convened a World OER Congress and released a 2012 Paris OER Declaration, which included a call for governments to 'encourage the open licensing of educational materials produced with public funds.'[17] UNESCO will be convening a second World OER Congress in Slovenia in 2017 to establish a 'normative instrument on OER.' OECD recently released its 2015 report: 'Open Educational Resources: A Catalyst for Innovation'[18] provides policy options to governments such as: 'Regulate that all publically funded materials should be OER by default. Alternatively, the regulation could state that new educational resources should be based on existing OER, where possible ("reuse first" principle).'[19]

As governments and foundations move to require the products of their grants and/or contracts be openly licensed, the implementation stage of these policies critical; open licensing policies should have systems in place to ensure that grantees comply with the policy, properly apply an open license to their work, and share an editable, accessible version of the OER in a public OER repository.[20]

A good example of an open education licensing policy done well is the US Department of Labor's 2010 Trade Adjustment Assistance Community College and Career Training Grant Program (TAACCCT) which committed US$2 billion in federal grant funding over four years to 'expand and improve their ability to deliver education and career training programs' (p.1). The intellectual property section of the grant program description requires that all educational materials created with grant funding be licensed under the Creative Commons Attribution (CC BY) license, and the Department required its grantees to deposit editable copies of the CC BY OER into skillscommons.org – a public open education repository.

A number of other nations, provinces and states have also adopted or announced open education policies relating to the creation, review, remix and/or adoption of OER. The Open Policy Registry[21] lists over 130 national, state, province, and institutional policies relating to OER, including policies like a national open licensing framework and a policy explicitly permitting public school teachers to share materials they create in the course of their employment under a CC license.

New open policy projects like the Open Policy Network[22] and the Institute for Open Leadership[23] are well positioned to foster the creation, adoption, and implementation of open policies and practices that advance the public good by supporting open policy advocates, organizations, and policy makers, connecting open policy opportunities with assistance, and sharing open policy information.

Because the bulk of education and research funding comes from taxpayer dollars, it is essential to create, adopt and implement open education licensing policies. The traditional model of academic research publishing borders on scandalous. Every year, hundreds of billions in research and data are funded by the public through government grants, and then acquired at no cost by publishers who do not compensate a single author or peer reviewer, acquire all copyright rights, and then sell access to the publicly funded research back to the University and Colleges. In the US, the combined value of government, non-profit, and university-funded research in 2013 was over US$158 billion[24] — about a third of all the R&D in the United States that year.

As governments move to require open licensing policies, hundreds of billions of dollars of education and research resources will be freely and legally available to the public that paid for them. Every taxpayer – in every country – has a reasonable expectation of access to educational materials and research products whose creation tax dollars supported.

Conclusion

If we want OER to go mainstream; if we want a complete set of curated OER for all grade levels, in all subjects, in all languages, customized to meet local needs; if we want significant funding available for the creation, adoption and continuous updating of OER – then we need (1) universal awareness of and systematic support for open educational resources and (2) broad adoption of open education licensing policies. When all educators are passionate about free and open access to their educational resources, when we change the rules on the money, when the default on all publicly funded educational resources is 'open' and not 'closed,' we will live in a world where everyone can attain all the education they desire.

Notes

[1] Hewlett Foundation: Open Educational Resources page: http://www.hewlett.org/programs/education/open-educational-resources.

[2] Most OER are 'born' digital, thought OER can be made available to students in both digital and printed formats. Of course, digital OER are easier to share, modify, and redistribute, but being digital is not what makes something an OER or not.

[3] Trans-Pacific Partnership Would Harm User Rights and the Commons: https://creativecommons.org/campaigns/trans-pacific-partnership-would-harm-user-rights-and-the-commons/

[4] Washington Post: Supreme Court: Copyright can be extended to foreign works once in public domain. Robert Barnes: https://www.washingtonpost.com/politics/supreme-court-copyright-can-be-extended-to-foreign-works-once-in-public-domain/2012/01/18/gIQAbqbr8P_story.html.

[5] https://www.opencontent.org/definition/.

[6] Game Changers: Chapter 6: Why Openness in Education? https://library.educause.edu/resources/2012/5/chapter-6-why-openness-in-education.

[7] For a short history of Creative Commons see: https://creativecommons.org/about/history/; for a full history on CC read: Viral Spiral – How the Commoners Built a Digital Republic of Their Own – David Bollier: http://bollier.org/viral-spiral-how-commoners-built-digital-republic-their-own.

[8] 2015 State of the Commons report: https://stateof.creativecommons.org/2015/.

[9] Viral Spiral – How the Commoners Built a Digital Republic of Their Own – David Bollier: http://bollier.org/viral-spiral-how-commoners-built-digital-republic-their-own.

[10] Note that Creative Commons (CC) licenses that include an ND clause (i.e., no derivatives) are not considered OER. For more information about CC licenses see: https://creativecommons.org/licenses/. For information about Open Source Initiative-approved licenses for software, see: https://open source.org/licenses.

[11] https://creativecommons.org/licenses.

[12] http://albertaoer.com.

[13] https://bccampus.ca/open-textbook-project.

[14] Digital rights management (DRM) schemes are used to restrict access to and use and/or modification of copyrighted works.

[15] For a full description of all four types of open education policies, see: Nicole Allen and Nick Shockey's 2014 Open Education Conference paper: Open Educational Resources and Public Policy: Overview and Opportunities http://conference.oeconsortium.org/2014/wp-content/uploads/2014/02/Paper_59-Policy.pdf.

[16] Game Changers: Chapter 6: Why Openness in Education? https://library.educause.edu/resources/2012/5/chapter-6-why-openness-in-education.

[17] 2012 Paris OER Declaration: http://www.unesco.org/new/en/communication-and-information/events/calendar-of-events/events-websites/World-Open-Educational-Resources-Congress.

[18] OECD Open Educational Resources: A Catalyst for Innovation: http://www.oecd.org/edu/open-educational-resources-9789264247543-en.htm.

[19] Page 131.

[20] For more detail on what governments should consider when implementing an open education licensing policy, see CCs 'Open Licensing Policy Toolkit' https://blog.creativecommons.org/2015/09/22/open-licensing-policy-toolkit-draft/.

[21] http://oerpolicies.org.

[22] https://openpolicynetwork.org.

[23] https://openpolicynetwork.org/iol.

[24] http://www.nsf.gov/statistics/2015/nsf15330/.

Openness and the Transformation of Education and Schooling

William G. Huitt* and David M. Monetti†

*Valdosta State University and Capella University, whuitt@valdosta.edu

†Valdosta State University

Editors' Commentary

It is tempting to couch the tension between open education and traditional educational models in simplistic terms such as free vs. commercial or permitting revision vs. static. In this chapter, authors Huitt and Monetti provide a more sophisticated discussion of the ways that openness is transforming education. They urge readers to consider the purpose and focus of education, outcomes and assessments, processes, and transparency. Parsing formal education into these components allows us more specific test cases to apply our thinking about open and to consider the transformation of schooling.

Education, at all levels and in its many forms, is experiencing significant social and economic pressure to change. There are many ideas about the source of this pressure, including:

- A recognition that the world is becoming increasing digital and global.[1]
- An increased importance of information and conceptual understanding.[2]
- A sociocultural context changing from an agricultural/industrial era focused on empire building to one of global, planetary collaboration.[3]
- An increased importance on creativity and innovation.[4]

How to cite this book chapter:

Huitt, W G, and Monetti, D M. 2017. Openness and the Transformation of Education and Schooling. In: Jhangiani, R S and Biswas-Diener, R. (eds.) *Open: The Philosophy and Practices that are Revolutionizing Education and Science.* Pp. 43–65. London: Ubiquity Press. DOI: https://doi.org/10.5334/bbc.d. License: CC-BY 4.0

Openness has been proposed as an important concept to address many of these concerns.[5]

One of the challenges in discussing openness in education and schooling is that the terms 'open' and 'education' are relatively complex. Firstly, education can refer to formal, informal, and non-formal aspects of teaching and learning. Additionally, education can refer to activities across the lifespan, from infancy and early childhood, to elementary, middle, and secondary school, to higher education, as well as adult education. Likewise, open can be used in a number of ways, from the aims and goals of education, to resources used, to the organizational structure of educational institutions. The purpose of this chapter is to clarify important dimensions that differentiate traditional and open education, to discuss our personal experiences with these, and provide our views as to next steps in the development of open education.

Types of Education

More often than not, when the term education is used, it refers to the formal organization of teaching and learning experiences for children and youth. However, that is only one aspect of education. La Belle (1982) advocated that a holistic, lifelong learning approach to human development and learning should consider all aspects of education, not just its formal conditions. This is especially important when considering open education as many of the dimensions discussed may apply more to the non-formal and informal types than to the formal.

- Formal education refers to any systematic form of teaching and learning that involves programs of study with defined, expected, and measurable outcomes. This, of course, would apply to early childhood through secondary schooling, but would also apply to higher education, including professional degree programs.
- Non-formal education refers to intentional teaching-learning experiences that do not necessarily involve multiple courses and measurable outcomes. This might involve extra-curricular activities for children and youth, but could also apply to workshops, training sessions, or other adult activities.
- Informal education refers to those spontaneous teaching-learning experiences that might occur as part of a classroom experience which result in unintended learning or brief encounters outside the classroom. It would also refer to most parent-child, workplace, or social interactions that are brief, natural encounters, incorporating some sort of teaching-learning experience.

The major point is that these three types of education can occur throughout the lifespan, although the relative mix would be different as the individual moves from infancy and early childhood through childhood and adolescence and into emerging, young, middle, and older adulthood. Therefore, when discussing open education, one must be mindful of these alternative contexts.

Openness

The terms 'open' and 'openness' have been used in a variety of ways when referring to education and schooling. In many cases, these terms are used to differentiate approaches to teaching and learning from more 'traditional' approaches. Table 1 provides an overview of some of the dimensions that have been used to distinguish between traditional and open educational experiences.

It is important to recognize that all aspects of education take place in a cultural milieu. For over 100 years, there has been a debate between those who advocate a more top-down, community-oriented approach to formal education (labeled here as traditional) and a more bottom-up, individualized approach (labeled here as open). Moreover, Berliner (1993) points to a debate as to what aspect of the community the curriculum should emphasize—the workplace (traditional) or living in a democratic society (open). This issue will be discussed more in the section on the purpose of education. Suffice it to say that in its present form, the debate is influenced heavily by the digital revolution that

	Traditional	Open
Transparency	Opaque or hidden data and decision making processes	Transparent data and decision making processes
Purpose	Socializing for factory work	Socializing for global democracy
Focus	Curriculum-centered	Person-centered
Desired Outcomes	Cognitive	Holistic
Assessment	Discrete cognitive knowledge	Authentic, holistic profile
Teaching Processes	Standardized, directed learning	Varied, as appropriate, with more self-regulated learning
Learning Tasks	Curriculum-directed	Problem- and project-based
Resources	Private enterprise controlled	Free or inexpensive
Work environment	Compartmentalized	Connected
Organizational structure	Centralized	Decentralized

Table 1: Analysis of Traditional and Open Education.

is permeating every aspect of both the lives of children and adolescents and the lives of the parents, teachers, and other adults in their communities.

Transparency

Transparency is one of the most important attributes of an open approach to teaching and learning. In fact, the phrase 'open and transparent' is likely a more correct description of openness as described in this chapter than is 'open and distance.' That is because open and transparent are both value-laded adjectives better contrasted with standardized and opaque whereas distance is better contrasted with face-to-face.

One of the ways that all forms of education can become more transparent is the connection among the various dimensions described in Table 1. For example, if a mission statement declares that open-mindedness is part of the mission of an educational program, a transparent organization would point to specific assessments, teaching processes, or learning tasks that directly support that statement. Absent that, the institution is implying that it does not want to be held accountable for actually achieving that purpose or goal. Allowing for free sharing of ideas and openly reporting outcomes and results of activities are behaviors that are consistent with open education. Education becomes more transparent by allowing stakeholders to easily be able to see its decision making and strategic planning processes.

Additional aspects related to the importance of transparency for open education will be highlighted throughout the remainder of the chapter.

Purpose

The debate as to the purpose of education in its many forms has been a source of contention from at least the ancient Greeks,[6] through the beginning of mass education[7] and the beginning of the industrial age in the United States,[8] to the transition to the modern era.[9] Historically, the focus on basic skills and standardized assessment is an aberration from the more generalized approach to developing the whole person. The emphasis on efficiency and preparation for factory work, while perhaps necessary for an industrial-age economy, is certainly not appropriate for a global, digital, information-age lifestyle.

In our view, a major purpose of education for children and youth should focus on developing the knowledge, attitudes, and skills necessary for global citizenship.[10] By this it is meant that children and youth should have the foundational skills sufficient to live and work anywhere in the world, in any lifestyle, that they may choose. At the same time, they should have the foundation to work with others to develop the neighborhoods and communities that will entice global citizens to live there. It is these two issues, an individual emphasis on personal freedom and a social emphasis on creating inviting communities that ought to be the focus for educating children and youth.

The purpose becomes a bit more diversified for higher education, as it ranges from technical schools and some community college programs preparing individuals for specific jobs to the continued general preparation of arts and science programs in colleges and universities to career preparation and advancement in fields such as business, education, health care, and the legal profession. Adult education has an even wider range from basic skills development to continuing education for professionals.

The importance of purpose cannot be overestimated. Several hundred years ago most people were farmers or serfs; only a small elite needed to have an ability to read and write. The industrial age brought a widening of job roles, requiring a minimal education for most people. In today's environment, the diversity of work and career options as well as lifestyles is changing so rapidly that some type of formal education will likely be required throughout an individual's life.[11] Therefore, a discussion of purpose must be one aspect for all strategic planning activities.[12]

Focus

Deciding on a general purpose for the various forms of education is only the first step. It is then necessary to consider how to make that broad statement more specific. For example, the traditional approach to early childhood to secondary schooling is to focus on the development of basic academic skills. The assumption is that if students have developed the academic competencies as described by the standards, they will be minimally prepared for successful adulthood in the twenty-first century. Likewise, the purpose of specific arts and science programs such as those found in the behavioral and social sciences is to provide opportunities for students to develop basic skills in scientific research, as well as concepts and principles that will allow them both to be successful in a wide variety of occupations and/or further specialized study.

Unfortunately, when learners only master traditional academic standards it might prepare them to be successful in higher education or advanced study, but that might not be sufficient to be successful in the workforce.[13] An open approach to formal schooling at the elementary, secondary, and tertiary levels would have a broader, more person-centered focus and would include flourishing and wellbeing; these latter are mainstays of the positive psychology movement.[14] More often than not, this broader approach includes a wider range of cognitive skills such as those involved in metacognition and problem-solving as well as competence in other domains such as emotional, social, and moral character development. Our personal experiences suggest this more open approach is not only possible, but it contributes to the development of more traditional academic skills.

We believe the same dichotomy can be seen in adult education. Whereas traditional adult education has focused on specific work-related skills, a more open approach would focus as much on developing the potential of the individual so as to empowering the person to take more control over his or her life. This would involve a consideration for how an individual forms and maintains professional relationships and how those could be mutually beneficial.

Desired outcomes

In the United States, the desired outcomes for children and adolescents are currently embodied in the Common Core State Standards[15]. This is consistent with a very narrow cognitive focus of human potential. However, theorists and practitioners have come to the realization that a free-flowing, dynamic environment with a corresponding exponential rate of change[16] requires a substantial change in desired outcomes of learning as well as the structure and processes of schooling and education.[17]

Whereas the industrial age required commonality in the development and use of basic academic skills as well as attitudes such as recognizing a supervisor's authority and a willingness to work on monotonous tasks,[18] the postmodern digital, information/conceptual age requires a much wider range of knowledge and skills.[19] While basic academic skills are still important, the ability to engage in such activities as group-based problem finding and problem solving; planning and implementing personally developed solutions that relate to personal interests and strengths; behaving in a morally and ethical manner; and engaging in meeting the perceived needs of the community and society are just as important.[20]

Unfortunately, there is less agreement about this broader set of knowledge, attitudes, and skills. Based on an analysis of recommendations from such researchers as Costa and Kallick (2000), Diener and Biswas-Diener (2008), the International Baccalaureate Organization (2013), Narvaez (2008), Partnership for 21st Century Skills (2009), and Seligman (2011), Huitt (2012) compiled a set of recommendations for desired learner outcomes. For example, in addition to those stated previously, some researchers have discussed the importance of open-mindedness and risk-taking, self-efficacy, resilience, and self-regulation. One of the ways that the behavioral and social science can contribute to this discussion is the development of instruments and methods that will allow these additional desired outcomes to be assessed in a reliable and valid manner.

Assessment

Of all the issues discussed so far, none is more important than the topic of assessment. That is because assessment embodies the purpose, focus, and desired outcomes of education and schooling and influences the creation of learning environments, teaching and learning processes, and learning tasks that will be used to facilitate that development of the desired outcomes. Hummel and Huitt (1994) used the acronym WYMIWYG (What You Measure Is What You Get) to describe this phenomenon. In fact, one might go so far as to suggest that if desired outcomes are not assessed, they are not really desired outcomes. The focus will be easily replaced by what seems to be more urgent, but ultimately less important, activities.

A traditional approach to assessment relies on standardized tests. This is seen not only in a traditional schooling environment, but also in a wide range of adult education for purposes of credentialing and promotion. This focus on cognitive knowledge is in spite of research showing that as much as two-thirds of the variance in adult success can be attributed to non-cognitive attributes.[21]

Developing appropriate assessments should be a priority of an open education approach to education and schooling. These will likely be norm-based assessments because developmentally-appropriate standards have not been established for these types of data-collection procedures.[22] This is exactly the approach being taken by the Collaborative for Academic, Social, and Emotional Learning (CASEL)[23] and Transforming Education[24] in their work with school districts. Gabrieli, Ansel, and Krachman (2015) provide evidence that focusing on these non-cognitive domains improves traditional academic learning. It is also a critical part of the positive psychology approach taken by Seligman and his colleagues[25] in their work with schools and Diener and his colleagues in their work with adults.[26]

One approach to assessment takes the form of e-portfolios such as those based on the domains of the Brilliant Star framework.[27] In this process, learners can upload digitized forms of artifacts (written documents, pictures, videos, etc.) that represent various levels of mastery in different domains. Learners can continue to add to this record throughout their childhood and adolescence and can use exemplars from their school-based extra-curricular activities as well as whatever else they may be doing. This is the same process that master artists and craftsmen use to display their work. An e-portfolio is a much more authentic process than any single measure that might be obtained using paper-pencil methods.

E-portfolios have been used successfully in psychology programs as a way for students to document and reflect on their learning.[28] The American Psychological Association (2013) developed a set of learning goals and objectives that could serve as a foundation for what should be assessed. One recommendation is to include some documentation as to how the learners applied their knowledge of the discipline to themselves such as their learning and cognitive styles, their strengths, and their personalities. We would also recommend that learners document their knowledge and application of physical wellbeing and their awareness of their own and others' emotions and their emotional self-regulation. These are some of the desired outcomes discussed by researchers cited above that are not included in the American Psychological Association (APA) guidelines.

Unfortunately, in prior progressive and education movements, while the advocated learning outcomes changed, the means for assessing learning did not. This was especially true for the 1960s version of open education in that learning assessments were not personalized, which was an important element of the various programs. Instead, the programs were required to use assessment methods more appropriate for an industrialized approach to teaching and learning – the use of standardized tests of basic skills for children and adolescents and standard assessments in higher education. This continues to be a challenge in the current open education movement.[29]

One final note on assessment. While it is readily acknowledged that feedback following action is necessary for learning, exponential learning is produced when both the action and feedback are shared among learners.[30] And when what is being shared has been digitalized, it can be shared at the speed of the internet, which is increasing exponentially.[31] This is why e-portfolios, when properly constructed, can impact learning in ways that traditional methods of assessment never could.

Teaching processes

Once it is accepted that assessments of learning will be standardized, standardizing teaching processes is the next logical step. Traditionally, this has meant that the method of choice is directed or explicit instruction with the teacher as the focus.[32] A more open approach would focus on methods that would emphasize self-regulated, lifelong learning.[33] This dichotomy is sometimes referred to as the 'Sage on the Stage' versus the 'Guide on the Side.'

However, it is important to realize that the issue may be more complex than this simple dichotomy would propose. For example, Gage and Berliner (1991) identified five principles that were adopted by those promoting open education in the 1960s and 1970s:

- Students will learn best what they want and need to know.
- Knowing how to learn is more important than acquiring a lot of knowledge.
- Self-evaluation is the only meaningful evaluation of a student's work.
- Feelings are as important as facts.
- Students learn best in a non-threatening environment.

These turned out to be incorrect principles when the desired outcomes were improved academic achievement, achievement motivation, locus of control, or self-concept although learners did show improved cooperativeness, creativity, independence, and positive attitudes toward school.[34] However, the importance of attention to affect as described by Rogers' and Freiberg's (1994) Facilitative

Teaching was shown to be a more relevant principle. In an assessment of educator's implementation of Rogers' and Freiberg's recommendations (in both open and traditional settings), Aspy and Roebuck (1975, 1977) found students performed better in school and had higher levels of self-concept when teachers:

- Responded to student feelings.
- Used student ideas in ongoing instructional interactions.
- Had more discussions with students (engaged in authentic dialogue).
- Praised students appropriately.
- Engaged in more authentic (less ritualistic) talk.
- Tailored content to individual's frame of reference.
- Smiled at students.

Those promoting a more open approach to teaching processes must be careful to identify the most relevant principles. A cautious, exploratory approach is certainly warranted.

In our experience, it is necessary to consider expectations of all stakeholders when advocating a change from a traditional to an open approach to education and schooling. For example, those in elementary and secondary schools must consider parental expectations as one of the major challenges to moving to a more varied approach to instruction. Parents' experiences with the more traditional model leads them to believe that direct instruction is the most appropriate method for teaching their children. If there is to be a successful transition to a more open approach, parent education must be a part of the process. They must be provided with opportunities to experience the efficacy of using a variety of teaching methods.

Likewise, those in higher education, especially in the liberal arts disciplines such as psychology, must consider the expectations of faculty and administrators with respect to promotion, tenure, and university ranking. While university faculty are evaluated in terms of research, teaching, and service, more often than not research and publications play a larger role than the other two. Even community college faculty are beginning to be expected to publish in the area of the teaching of their discipline.[35] However, the level of innovation in teaching practice is not one of the criteria normally used to evaluate faculty and departments. Yet the requirement for innovative practice in teaching must be addressed if higher education is to keep pace with the disruptive sociocultural change in which it is embedded.

In particular, the processes of learning and their associated teaching methods must be made known to all important stakeholders. For example, it can be shown that current methods such as the flipped classroom, problem-based learning, and project-based learning are all supported by learning theories.[36] By showing stakeholders that next practices are simply a rearrangement or extension of best practices, they are more likely to be supportive of previously unfamiliar methods. The key is making these approaches transparent so that everyone understands educators are not making changes simply for change sake.

Learning tasks

There is no greater difference between traditional and open approaches to education and schooling than in the description of learning tasks. Even though other methods such as concept mapping and cooperative learning are used in traditional classrooms, direct instruction is still the dominant method used in the United States and throughout the world.[37] This means that learners spend large amounts of learning time listening to or watching a teacher and engaged in practicing isolated tasks that are directly related to curricular objectives. This is in spite of research showing that reciprocal teaching (where learners take responsibility for teaching other learners), the use of meta-cognitive strategies, and student self-verbalization or self-questioning all explain more variance in test scores than does the use of direct instruction.[38]

One result of having a more open approach to describing desired outcomes and its subsequent impact on a wider set of assessments is that a wider range of learning tasks will be necessary to accomplish those. For example, once different aspects of emotional and social development are deemed important, it is then necessary to create learning tasks that will allow learners to develop those competencies. The same is true for self-regulation, moral character, or any number of other desired objectives.

The most important principle is that learning tasks should address a wider range of desired outcomes and those outcomes must be appropriately assessed. The creation or selection of learning tasks that specifically address desired outcomes must be designed in such a way that assessment FOR learning is designed into each learning task.[39] For example, when a small group is involved in a discussion or collaborative learning activity, other students could be assigned as observers. The student observers collect data on desired competencies for working in groups and that data is shared with those in the discussant group. As students become more skilled as observers, they will become more aware of the competencies they should be developing when they are in a discussion group.

Experiential learning, especially academic service learning, should be part of every school curriculum. Rogers and Freiberg (1994) showed that experiential learning provides learners with an opportunity to know-how in addition to know-what, which makes the learning experience more personally significant. This is especially true for academic service learning as students are able to see a direct purpose for academic learning that makes academic learning meaningful.[40]

One of us (WH) is currently working with colleagues to develop a series of undergraduate courses that will provide learners with guided experiences at multiple levels of community development.[41] A central concept is that young people need to have a variety of experiences that will allow them to make better decisions about how they want to contribute to the development of a society in which they would like to live. Without these types of learning experiences, young people, for the most part, are only guessing as to what their interests and strengths might be and how those could be used for social good. Providing learners with authentic learning tasks with built-in opportunities for feedback and reflection should be given a high priority in an open education approach to teaching and learning.

Resources

Open access to resources is probably one of the most acknowledged aspects of an open education movement. This advocacy of free or inexpensive access to important information[42] is in direct contrast with a traditional approach where resources are controlled by for-profit corporations or professional organizations. Fortunately, there is an exponential growth in materials that are either free or relatively inexpensive; these are extensively covered in other chapters in this book (e.g., Chapters 17 & 18). Open access to resources is a central pillar of a more open approach to education and schooling.

One of us (WH) has been involved in the process of producing and sharing free resources for the purposes of teaching education and psychology courses since the early 1990s. The materials on the website[43] have been used to create a number of courses whose materials are largely comprised of free resources. It is our expectation that this trend will continue, and even accelerate, in the near future.

Work environment

Creating a more open work environment is one area where educational institutions could learn from their counterparts in private enterprise. High-tech organizations such as Google are well-known for their willingness to break down compartmentalization and create more inter-departmental and connected communication systems.[44] For the most part, educational institutions are still organized via academic departments with very little cross-fertilization.

One area where this tradition is being challenged in elementary and secondary education is in the area of STEAM (science, technology, engineering, arts, mathematics) projects. With some projects including the social sciences as well as natural sciences in the projects, they are leading the way in creating a more connected work environment. For example, High Tech High School regularly integrates the arts in its project-based instructional program.[45] In fact, entire school districts are now coordinating their efforts to create an integrated curriculum.[46]

The Character through the Arts project[47] is another example of work on which one of us (WH) contributed. The focus of this project was on the development of arts-integrated units for elementary and middle schools.[48] One of the highlights of the project was the collaboration of one of the participating schools with the theater department at the local university.[49] Through this collaboration, the entire middle and upper schools participated in two one-day events that allowed students to explore important character issues such as understanding the consequences of one's decisions and the burdens (through the study of Macbeth) and responsibilities of leadership (through the study of Antigone).

Behavioral and social science departments could provide leadership in creating STEAM-oriented case studies and projects that would have learners connect across multiple disciplines. Lisa Delissio's blog[50] provides extensive examples of current work in this area. When these engage learners in experiential education, especially service learning activities that benefit the local community, the learning and work environment can better address the need for the working environment to contribute to the more open, holistic desired outcomes, assessments, teaching processes, and learning tasks described above.[51] These types of experiences make academic learning more relevant and meaningful to learners and will be a pillar in the transition away from the centuries-old approach to compartmentalizing learning experiences and preparing learners for the work of the future.[52]

Organizational structure

The hierarchical structure that presently dominates educational institutions will slowly give way to more holarchical structures that rely more on consultation than authority.[53] The key element is that the decision making process moves from an industrial-age, military-like centralized decision-making process to one more like a set of embedded networks where most decisions are made by those who will actually implement them. Ismail, Malone, and van Geest (2014) state quite explicitly that organizations using these holarchical types of structures will be able to outperform more traditionally-organized institutions in times of exponential change. These organizations are simply able to more quickly resolve conflicts among various stakeholders and get on with creating value for customers.

While we do not have any direct experience with holarchical organizational structures in our educational experiences, our experience in using some of the principles with non-profit and religious organizations has demonstrated the power of a more nimble, agile approach to decision making. The time it takes for all stakeholders to discuss an issue thoroughly is more than made up for when actually implementing the decision. The buy-in from those implementing the decision is enhanced because they had influence in its creation. Those involved in educational institutions at all levels will need to acquire the knowledge, attitudes, and skills to create and work in these more open organizational structures.

More Open Psychology Departments

Even though we will address psychology departments specifically, our thoughts and recommendations are applicable to all departments in the liberal arts, especially to those in the behavioral and social sciences or educational psychology departments in colleges of education. That is because higher education is experiencing significant economic pressure to change. Enrollment at many campuses is down and costs are up. State contributions to public higher education are in most cases either flat-lined or eroding. Endowments are down as well, in lockstep with recent global financial and market reductions. Student debt is increasing and families are forced to analyze the kind of return on investment that they can anticipate from a college education.

However, to look at higher education classrooms at institutions around the world, it is likely that one would see familiar rows that have come to typify the traditional face-to-face college experience. As we have discussed, this tradition will see continued challenge in an era where students expect immediacy and are increasingly comfortable with mobile computing and collaborating and sharing information through social networks.

One reaction to such challenges is often righteous indignation, where traditionalists argue that openness and technological innovation amount to little more than window dressing. This same line of thinking was once held by video rental operations (like Blockbuster) who failed to embrace the digital content revolution. At the same time, any innovation or modification made which is not at its core imbued with quality will also fail. People pay more for luxury items because their quality is typically higher than competitors. Providing quality academic learning experiences is one of the primary reasons why investments in faculty and students are a critical component to educational ratings. The public tends to skeptically observe institutions with slick marketing campaigns without corresponding investments in students and faculty.

We have discussed ten characteristics that can be utilized to reflect upon the degree of openness in a k-12 (kindergarten through secondary school) or higher education unit, like a psychology department. These components were: transparency, purpose, focus, desired outcomes, assessment, teaching processes, learning tasks, resources, work environment, and organizational structure. With these elements in mind, what might a more open psychology department look like?

One of the key factors toward moving to a more open department would be an assessment of the purpose and focus of psychology departments. Many academic units tend to emphasize a narrow component of an open curriculum.

Moreover, the curriculum is generally thought of as fixed and does not routinely change based on the students in the class. In fact, the pacing and sequence of instruction is often decided before the professor meets the learners. One change that could be made is for faculty to assess students about their knowledge, attitudes, and skills as they relate to a more open set of desired outcomes. Faculty in a more open department will attempt to focus more on student strengths in terms of their interests and strengths. Case studies and projects would focus on these and learners would have the opportunity to make choices about how they would apply academic concepts and principles taught in the course.

Open psychology departments will also embrace that idea that a percentage of students (perhaps a vast majority for some institutions) will decide not to immediately attend graduate school. The challenge will be to help those students connect what they have learned about psychological science to their careers. There is a reason so many students select psychology as a discipline – it fosters curiosity, objectivity, awareness of the impact of bias, and skills in both writing and quantitative data analysis. Open departments will continue to expand the ways that they have connected students to the world of work and sharing these reasons with employers. Helping students participate in service learning, internships, relevant case studies, and capstone projects will help demonstrate to employers the intellectual, emotional, and social skills graduates are able to bring to the workplace.

Another area where openness can occur is around how data is collected and utilized in departments. Currently data is gathered primarily for the purpose of grading and accreditation. Efforts to organize data collection and analysis through reflective e-Portfolios using APA's guidelines are a step in the right direction.[54] Assisting learners to develop compilations, such as competency profiles on LinkedIn, is one way that individuals can connect personal learning to work-related social networking.

Another step would be to make data even more transparent to stakeholders such as students, parents, and citizens. The decisions that individuals make regarding higher education would be enhanced by providing aggregated information on these desired outcomes. Asking students to voluntarily share their e-Portfolios and competency profiles could become a dominant form of departmental accountability. Our experience with e-portfolios shows they are an excellent method for getting a holistic overview of an individual's learning and development, and competency profiles can provide a similar, though somewhat briefer, role.

Instructional processes have already starting shifting to more open practices. There has been a growth of delivery methods that include face-to-face, fully online, and blended approaches. Even face-to-face instruction has changed with increased use of flipped classrooms and problem- and project-based learning. These approaches have given students increased opportunities to develop a more holistic set of desired outcomes.

Options within degree programs are also starting to expand. In addition to the standard psychology curriculum with some course choices, varied options are becoming more prevalent. For students where cost is a very large factor and potential for college dropout is high, institutions are experimenting with direct path degree programs which minimize advising and course selection errors. There are also more options where students basically create their own degree program which encourages interdisciplinary thinking and learning.

Open departments will do a better job of intervening to help students avoid failing out. How students are grouped is a powerful intervention. Psychology departments tend to be really large; majors could be separated into smaller groups of learning communities and focus on helping students become connected to a much smaller and tighter peer group.

Another open practice would be to ensure that faculty meetings are places were data are discussed. In departmental meetings, faculty often do not have the hard data regarding the functioning of the department with which to start a discussion of improvement. Data could bring light to many important bottlenecks within the psychology curriculum. For example, data could help faculty to more empirically answer questions related to the connection of desired outcomes, teaching practices, and learning experiences. Faculty would be encouraged to share examples of innovations that did not work as designed. Failing often and quickly is the hallmark of the most successful innovative institutions.

Summary and Conclusions

In summary, while it is generally recognized that the global sociocultural environment, heavily influenced by digitization and exponential change, is a current reality, it is less obvious as to how education and schooling need to be modified as a result of these changes. At the very least it requires a change in desired outcomes and related changes in assessments. Changing the first (e.g., a focus on higher levels of critical and/or creative thinking) without changing the second (e.g., creating new ways of assessing those more varied outcomes) will result in continued use of instructional methods more suited to the industrial age than the digital, information age.[55] At the same time, the organizational structure of schooling and education needs to be changed so that it is less hierarchical, more open, and more transparent. This is beginning to change with an increase in magnet and charter schools for elementary and secondary learners[56] as well as rapidly growing alternatives in distance learning for higher education[57] and adult education.[58]

The positive psychology initiative and a corresponding focus on creating learning experiences that result in higher levels of flourishing and wellbeing as referenced above demonstrate that all stakeholders can be provided with a solid knowledge base on how to create learning experiences that are increasingly relevant to children, youth, and adults. This does not mean a total abandonment of an emphasis on academic knowledge, but it does mean that society-wide discussions need to be held as to what aspects of the academic curriculum are absolutely needed for global citizenship, which should be taught just in case they are relevant, and which should be taught on a just-in-time basis so they can be used in a problem- or project-based learning experience.[59]

Finally, the phrase 'open and distance education' should be replaced with the more accurate phrase 'open and transparent education.' Distance education is merely a delivery system and is not inherently open or transparent. Our experience shows that distance education can be just as traditional as any face-to-face classroom. Distance education will become increasingly relevant to the extent that educators can address the teaching and learning processes and the means of assessment that are appropriate for a global, digital, information-rich environment of living and learning. It is open and transparent that is the key to future models of education and schooling, not the delivery system, as convenient as it might be. It will be very interesting to watch how open and transparent practices shape education. We hope that all interested stakeholders will participate in this important discussion.

Acknowledgements

We would like to thank Sheri Dressler and Marsha Huitt for assistance in developing this chapter.

Notes

[1] Prensky, 2010.
[2] Huitt, 2007.
[3] Gilman, 1993.
[4] Diamandis & Kotler, 2012, 2015; Wagner, 2012.
[5] Price, 2013.
[6] Smith, 2001.
[7] Filler, 1983.
[8] Taylor, 1913; Dewey, 1991, 1997.
[9] Freire, 2000, 2013; Wagner, 2012.
[10] Huitt, 2013.
[11] Gilman, 1993; Price, 2013.
[12] Delprino, 2013; Proctor, 1997.
[13] Gardner, 1995; Goleman, 1995; Sternberg et al., 1995.
[14] Diener & Biswas-Diener, 2008; Rogers & Freiberg, 1994; Seligman, 2011.
[15] Common Core State Standards, n.d.
[16] Kurzweil, 2004; Diamandis & Kotler, 2012, 2015.

17 Prensky, 2010; Tapscott, 2008.
18 Chance, as cited in Huitt, 1999.
19 Huitt, 2007.
20 Huitt, 2013.
21 Gardner, 1995; Goleman, 1995; Sternberg et al., 1995.
22 Huitt & Monetti, 2015.
23 CASEL, n.d.
24 Transforming Education, n.d.
25 Duckworth & Seligman, 2005; Seligman, 2008, 2012; Seligman et al., 2009.
26 Diener et al., 2010; Diener, Inglehart & Tay, 2013.
27 Brilliant Star Framework, n.d.
28 Birkett, Neff & Pieper, 2013, Stephens & Moore, 2006.
29 Admiraal, Huisman & Pilli, 2015; Caple & Bogle, 2013.
30 Diamandis & Kotler, 2012.
31 Diamandis & Kotler, 2015.
32 Huitt, Monetti & Hummel, 2009; Monetti, Huitt & Hummel, 2006.
33 Cleary & Zimmerman, 2004; Sitzman & Ely 2011.
34 Giaconia & Hedges, 1982.
35 Palmer, 2015.
36 Huitt & Vernon, 2015.
37 Huitt et al., 2009.
38 Huitt, Huitt, Monetti & Hummel, 2009.
39 Stiggins, 2008; Wiliam, 2014.
40 Warren, 2012.
41 Huitt & Dressler, 2014.
42 Downes, 2011; McKerlich, Ives & McGreal, 2013.
43 Educational Psychology Interactive , n.d.
44 Price, 2013.
45 High Tech High School, n.d.
46 Integrated Curriculum, n.d.
47 Character through the Arts, n,d,
48 Educational Psychology Interactive, n.d.
49 Huitt & Huitt, 2007.
50 Lisa Delissio, n.d.
51 Bowdon, 2013.
52 Ross, 2016.
53 Robertson, 2015.
54 American Psychological Association, 2013.
55 Hummel & Huitt, 1994.
56 Huitt, 2006.
57 Allen & Seaman, 2014.
58 Milana, 2012.
59 Huitt, 2013.

References

Admiraal, W., Huisman, B., & Pilli, O. (2015). Assessment in massive open online courses. *The Electronic Journal of e-Learning, 13*(4), 207–216. Retrieved from http://www.ejel.org/volume13/issue4

Allen, I. E., & Seaman, J. (2014, January). *Grade change: Tracking online education in the United States.* Babson Park, MA and San Francisco, CA: Babson College, Babson Survey Research Group and Quahog Research Group LLC. Retrieved from http://www.onlinelearningsurvey.com/reports/gradechange.pdf

American Psychological Association. (2013). *APA guidelines for the undergraduate psychology major.* Washington, DC: Author. Retrieved from http://www. apa.org/ed/precollege/about/psymajor-guidelines.aspx

Aspy, D., & Roebuck, F. (1975). The relationship of teacher-offered conditions of meaning to behaviors described by Flanders' interaction analysis. *Education 95*, 216–222.

Aspy, D, & Roebuck, F. (1977). *Kid's don't learn from people they don't like.* Amherst, MA: Human Resources Development Press.

Berliner, D. (1993). The 100-year journey of educational psychology: From interest, to disdain, to respect for practice. In T. Fagan & G. VandenBos (Eds.), *Exploring applied psychology: Origins and critical analysis.* Washington, DC: American Psychological Association.

Birkett, M., Neff, L., & Pieper, S. (2013, February 2). Portfolios in psychology classes: Dispelling myths. *Observer, 26*(2). Retrieved from http://www. psychologicalscience.org/index.php/publications/observer/2013/february-13/ portfolios-in-psychology-classes.html

Bowdon, M. (2013). *Engaging STEM in higher education: A faculty guide to service-learning.* Tallahassee, FL: Florida Compass Compact. Retrieved February 2016, from http://www.floridacompact.org/pdf/stem-manual.pdf

Brilliant Star Framework. (n.d.). Available at http://www.educpsyinteractive. org/katie

Caple, H., & Bogle, M. (2013). Making group assessment transparent: What wikis can contribute to collaborative projects. *Assessment & Evaluation in Higher Education, 38*(2), 198–210. DOI: https://doi.org/ 10.1080/02602938. 2011.618879

Character through the Arts. (n.d.). Available at http://www.cccoe.k12.ca.us/ edsvcs/stem.html see also http://www.pioneerresa.org/CTTA/main.html

Cleary, T., & Zimmerman, B. (2004). Self-regulation empowerment program: A school-based program to enhance self-regulation and self-motivated cycles of student learning. *Psychology in the Schools, 41*(5), 537–550.

Collaborative for Academic, Social, and Emotional Learning (CASEL). (n.d.) Available at http://www.casel.org/

Common Core State Standards. (n.d.). Available at http://www.corestandards.org/

Costa, A. L., & Kallick, B. (2000). *Habits of mind: A developmental series.* Alexandria, VA: Association for Supervision and Curriculum Development.

Delprino, R. (2013). *Human side of the strategic planning process in higher education*. Ann Arbor, MI: Society for College and University Planning.

Dewey, J. (1991). *School and society* and *The child and the curriculum* (Reissue edition). Chicago: University of Chicago Press. [Originally published in 1896 and 1902].

Dewey, J. (1997). *Experience and education*. New York: Macmillan. [Originally published in 1938].

Diamandis, P., & Kotler, S. (2012). *Abundance: The future is better than you think*. New York, NY: Free Press.

Diamandis, P., & Kotler, S. (2015). *Bold: How to go big, create wealth and impact the world*. New York, NY: Simon & Schuster.

Diener, E., & Biswas-Diener, R. (2008). *Happiness: Unlocking the mysteries of psychological wealth*. Malden, MA: Blackwell Publishing.

Diener, E., Inglehart, R., & Tay, L. (2013). Theory and validity of life satisfaction scales. *Social Indicators Research, 112*(3), 497–527. DOI: https://doi.org/10.1007/s11205-012-0076-y

Diener, E., Wirtz, D., Tov, W., Kim-Prieto, C., Choi, D., Oishi, S., & Biswas-Diener, R. (2010). New well-being measures: Short scales to assess flourishing and positive and negative feelings. *Social Indicators Research, 97*(2), 143–156. DOI: https://doi.org/10.1007/x11205-009-9493-y

Downes, S. (2011, July 14). Open education resources: A definition. [Web log comment]. Retrieved from http://halfanhour.blogspot.com/2011/07/open-educational-resources-definition.html

Duckworth, A., & Seligman, M. (2005). Self-discipline outdoes IQ in predicting academic performance in adolescents. *Psychological Science, 16*(12), 939–944.

Educational Psychology Interactive. (n.d.) Available at http://www.edpsycinteractive.org

Filler, J. W., Jr. (1983). Service models for handicapped infants. In S. G. Garwood & R.R. Fewell (Eds.), *Educating handicapped infants* (pp. 369–386). Rockville, MD: Aspen Systems.

Freire, P. (2000). *Pedagogy of the oppressed*. London, UK & New York, NY: Bloomsbury Academic. [Originally published in 1970]

Freire, P. (2013). *Education for critical consciousness*. London, UK & New York, NY: Bloomsbury Academic. [Originally published in 1974]

Gabrieli, C., Ansel, D., & Krachman, S. B. (2015, December). *Ready to be counted: The research case for education policy action on non-cognitive skills*. [Working Paper]. Boston, MA: Transforming Education. Retrieved from http://static1.squarespace.com/static/55bb6b62e4b00dce923f1666/t/5665e1c30e4c114d99b28889/1449517507245/

Gage, N., & Berliner, D. (1991). *Educational psychology* (5th ed.). Boston, MA: Houghton, Mifflin.

Gardner, H. (1995). Cracking open the IQ box. In S. Fraser (Ed.), *The bell curve wars* (pp. 23–35). New York: Basic Books.

Goleman, D. (1995). *Emotional intelligence*. New York, NY: Bantam Books.

Gianconia, R., & Hedges, L. (1982). Identifying features of effective open education. *Review of Educational Research, 52*(4), 579–602.

Gilman, R. (1993, Fall). What time is it? Finding our place in history. *In Context, 36.* Retrieved from http://www.context.org/iclib/ic36/gilman1/

High Tech High School. (n.d.). Available at http://www.hightechhigh.org/projects

Huitt, W. (1999). *Success in the information age: A paradigm shift.* Revision of paper developed for a workshop presentation at the Georgia Independent School Association, Atlanta, Georgia, November 6, 1995. Retrieved from http://www.edpsycinteractive.org/papers/infoage.pdf

Huitt, W. (2006, April 25). *Educational accountability in an era of global decentralization.* Paper presented at the International Networking for Educational Transformation (iNet) Conference, Augusta, GA. Retrieved from http://www.edpsycinteractive.org/papers/edaccount.pdf

Huitt, W. (2007). *Success in the Conceptual Age: Another paradigm shift.* Paper presented at the 32nd Annual Meeting of the Georgia Educational Research Association, Savannah, GA, October 26. Retrieved from http://www.edpsycinteractive.org/papers/conceptual-age-s.pdf

Huitt, W. (2012, June). Developing the whole person: Profile domains with detailed descriptors. *Educational Psychology Interactive*. Valdosta, GA: Valdosta State University. Retrieved from http://edpsycinteractive.org/brilstar/CurrMap/ltr/drop-down-menu-template-new.pdf

Huitt, W. (2013). Curriculum for global citizenship. *International Schools Journal, 33*(1), 76–81. Retrieved from http://www.cosmic-citizenship.org/2013-huitt-curriculum-for-global-citizenship.pdf

Huitt, W., & Dressler, S. (2014, August). *An integrative approach to community development.* Atlanta, GA: Community Development through Academic Service Learning. Retrieved from http://cd-asl.org/overview.html

Huitt, M., & Huitt, W. (2007). *Connecting literature and drama to develop critical and creative thinking.* Paper presented at the China–U.S. Conference on Literacy, Bejing, People's Republic of China, July 23–26. Retrieved from http://www.edpsycinteractive.org/papers/conlitdrama.doc

Huitt, W., Huitt, M., Monetti, D., & Hummel, J. (2009). *A systems-based synthesis of research related to improving students' academic performance.* Paper presented at the 3rd International City Break Conference sponsored by the Athens Institute for Education and Research (AITNER), Athens, Greece, October 16–19. Retrieved from http://www.edpsycinteractive.org/papers/improving-school-achievement.pdf

Huitt, W., & Monetti, D. (2015). Norm-based assessments. In J. Spector (Ed.), *The SAGE encyclopedia of educational technology* (pp. 545–547). Thousand Oaks, CA: SAGE Publications, Inc. DOI: https://doi.org/10.4135/9781483346397.n226

Huitt, W., Monetti, D., & Hummel, J. (2009). Designing direct instruction. Pre-publication version of chapter published in C. Reigeluth and A. Carr-Chellman, *Instructional-design theories and models: Volume III, Building a common knowledgebase* (pp. 73–97). Mahwah, NJ: Lawrence Erlbaum Associates. Retrieved from http://www.edpsycinteractive.org/papers/designing-direct-instruction.pdf

Huitt, W., & Vernon, K. (2015). *The flipped classroom and PBL: Theory and practice.* Presentation at the European Council for International Schools (ECIS), Barcelona, Spain, November 21. Retrieved from http://www.edpsycinteractive.org/edpsyppt/Presentations/flipped-classroom-and-pbl.html

Hummel, J., & Huitt, W. (1994, February). What you measure is what you get. *GaASCD Newsletter: The Reporter*, 10–11. Retrieved from http://www.edpsycinteractive.org/papers/wymiwyg.html

Integrated Curriculum. Available at http://www.cccoe.k12.ca.us/edsvcs/stem.html

International Baccalaureate Organization. (2013). The *IB learner profile.* Retrieved from http://www.ibo.org/globalassets/digital-tookit/flyers-and-artworks/learner-profile-en.pdf

Ismail, S., Malone, M., & van Geest, Y. (2014). *Exponential organizations: Why new organizations are ten times better, faster, and cheaper than yours (and what to do about it).* New York, NY: Diversion Books.

Kurzwiel, R. (2004). Kurzweil's law (aka "the law of accelerating returns"). [Web log comment]. Retrieved from http://www.kurzweilai.net/kurzweils-law-aka-the-law-of-accelerating-returns

La Belle, T. (1982). Formal, nonformal and informal education: A holistic perspective on lifelong learning. *International Review of Education, 28*(2), 159–175.

Lisa Delissio. (n.d.). Available at https://stemtosteamihe.wordpress.com/

McKerlich, R., Ives, C., & McGreal, R. (2013). Measuring use and creation of open educational resources in higher education. *The International Review of Research in Open and Distance Learning, 14*(4), 90–112.

Milana, M. (2012). Globalization, transnational policies and adult education. *International Review of Education, 58*(6), 777–797. DOI: https://doi.org/10.1007/s11159-012-9313-5

Monetti, D. M., Hummel, J. H., & Huitt, W. G. (2006). Educational psychology principles that contribute to effective teaching and learning. *International Journal of Arts & Sciences, 1*(1): 22–25.

Narvaez, D. (2008). Triune ethics: The neurobiological roots of our multiple moralities. *New Ideas in Psychology, 26*, 95–119. Retrieved from http://www.nd.edu/~dnarvaez/documents/TriuneEthicsTheory0725071.pdf

Partnership for 21st Century Skills. (2009). *P21 framework definitions.* Retrieved from http://www.p21.org/storage/documents/P21_Framework_Definitions.pdf

Palmer, S. (2015). *Toxic childhood: How the modern world is damaging our children and what we can do about it.* London, UK: Orion.

Prensky, M. (2010). *Teaching digital natives: Partnering for real learning.* Thousand Oaks, CA: Corwin.

Price, D. (2013). *Open: How we'll work, live and learn in the future.* London, UK: Crux Publishing.

Proctor, T. (1997). Establishing a strategic direction: A review. *Management Decision, 35*(2): 143–154.

Robertson, B. (2015). *Holacracy: The new management system for a rapidly changing world.* New York, NY: Henry Holt & Co.

Rogers, C., & Freiberg, H. J. (1994). *Freedom to learn* (3rd ed.). New York, NY: Merrill. [First edition published in 1969]

Ross, A. (2016). *Industries of the future.* New York, NY: Simon & Schuster.

Seligman, M. (2008). Positive education and the new prosperity: Australia's edge. *Education Today 8*(3), 20–21. Retrieved from http://www.minnis journals.com.au/articles/ET%20Aug-Sep%20web%20pp%2020_21%20%2810.9.08%29–11.pdf

Seligman, M. (2011). *Flourish: A visionary new understanding of happiness and well-being.* New York, NY: Free Press.

Seligman, M. [Video file]. *Wellbeing before learning: Flourishing students, successful schools.* Keynote address to the Wellbeing Before Learning: Flourishing students, successful schools conference in Adelaide, Feb 27. Retrieved from https://www.youtube.com/watch?v=Rl8yX_8LVnc&feature=youtu.be

Seligman, M., Ernst, R., Gillham, J., Reivich, K., & Linkins, M. (2009). Positive education: Positive psychology and classroom interventions. *Oxford Review of Education, 35*(3), 293–311.

Sitzman, T., & Ely, K. (2011). A meta-analysis of self-regulated learning in work-related training and educational attainment: What we know and where we need to go. *Psychological Bulletin, 127*(1), 421–442. DOI: https://doi.org/10.1037/a0022777

Smith, M. (2001). Aristotle and education. *Infed.* Retrieved from http://infed.org/mobi/aristotle-and-education/

Stephens, B. R., & Moore, D. (2006). Psychology and e-portfolios enhance learning, assessment, and career development. In A. Jafari, & C. Kaufman (Eds.) *Handbook of research on e-portfolios* (pp. 520–531). Hershey, PA: Idea Group Publishing. DOI: https://doi.org/10.4018/978-1-59140-890-1.ch046

Sternberg, R., Wagner, T., Williams, W., & Horvath, J. (1995). Testing common sense. *American Psychologist, 50*(11), 912–927.

Stiggins, R. (2008). *Assessment FOR learning, the achievement gap, and truly effective schools.* Presentation at the Educational Testing Service and College Board conference, Educational Testing in America: State Assessments, Achievement Gaps, National Policy and Innovations, Washington, DC, September 8. Retrieved from http://www.ets.org/Media/Conferences_and_Events/pdf/stiggins.pdf

Tapscott, D. (2008). *Grown up digital: How the net generation is changing the world*. New York, NY: McGraw-Hill.

Taylor, F. W. (1913). *The principles of scientific management*. New York, NY: Dover Publications. Retrieved from https://books.google.com/books?id= HoJMAAAAYAAJ

Transforming Education. (n.d.). Available at http://www.transformingeducation .org/

Wagner, T. (2012). *Creating innovators: The making of young people who will change the world*. New York, NY: Scribner.

Warren, J. (2012). Does service learning increase student learning?: A meta-analysis. *Michigan Journal of Community Service Learning, 19*(1), 69–73. Retrieved from http://quod.lib.umich.edu/cgi/p/pod/dod-idx/ does-service-learning-increase-student-learning-a-meta.pdf?c=mjcsl;i dno=3239521.0018.205

Wiliam, D. (2014). *Formative assessment and contingency in the regulation of learning processes*. Paper presented in a Symposium entitled Toward a Theory of Classroom Assessment as the Regulation of Learning at the annual meeting of the American Educational Research Association, Philadelphia, PA, April. Retrieved from http://dylanwiliam.org/Dylan_Wiliams_website/ Papers_files/Formative%20assessment%20and%20contingency%20in%20 the%20regulation%20of%20learning%20processes%20%28AERA%20 2014%29.docx

What Can OER Do for Me? Evaluating the Claims for OER

Martin Weller[*], Beatriz de los Arcos, Rob Farrow,
Rebecca Pitt and Patrick McAndrew

Institute of Educational Technology, The Open University
*Martin.Weller@open.ac.uk

Editors' Commentary

As advocates for OER describe its transformative potential, it is critical that these claims are scrutinized using empirical methods. In this chapter, the authors— all of whom are affiliated with the OER Research Hub at the Open University— provide an overview of their team's research on three types of OER users: informal learners, formal learners, and educators. In doing so they address a broad series of 11 hypotheses related to performance, openness, access, retention, reflection, finance, indicators, support, transition, policy, and assessment. The chapter concludes with a discussion of these findings and a call for research to play a central role as OER continues to go mainstream.

Introduction

This chapter will explore the role of research in the emerging OER discipline, with a particular focus on the work of the OER Research Hub at the UK Open University. OER have the potential to impact upon many aspects of education, such as improved performance, cost savings and development of new approaches to teaching. In this chapter we will explore some of the claims made

---

How to cite this book chapter:
Weller, M, de los Arcos, B, Farrow, R, Pitt, R, and McAndrew, P. 2017. What Can OER Do for Me? Evaluating the Claims for OER. In: Jhangiani, R S and Biswas-Diener, R. (eds.) *Open: The Philosophy and Practices that are Revolutionizing Education and Science.* Pp. 67–77. London: Ubiquity Press. DOI: https://doi.org/10.5334/bbc.e. License: CC-BY 4.0

for the benefit of OER, and the research that is now emerging to support these claims. The benefits of OER are often deemed to be self-evident – a free text-book is 'evidently' better for a student than one which costs US$100, for example but at the start of the OER movement this evidence was often lacking. As the field matures, the importance of research to demonstrate the actual impact of OER in expected, and unexpected, ways becomes useful in deciding how the can be most effectively deployed.

The start of the OER movement is often given as the announcement of the MIT OpenCourseWare project in 2001, although this itself grew out of other movements such as learning objects, open source and open universities.[1] However, if we accept 2001 as the start of the OER movement in earnest, then it is now some 15 years old. This is still relatively young in terms of academic disciplines, but it provides a useful timeframe to examine how research has changed over that period.

In terms of establishing itself as a global movement, OER has been something of a success story compared with many educational developments. There are repositories in most major languages, considerable funding has been provided by foundations such as Hewlett, and national bodies such as JISC (the Joint Information Systems Committee) in the UK, and there have been a number of policies at the national, and institutional level setting out a programme for releasing or funding OER.[2]

The advent of open textbooks is an example of an area that has seen particularly clear gains. These are electronic versions of standard textbooks that are freely available and can be modified by users. The physical versions of such books are available at a low cost to cover printing, for as little as US$5.[3] The motivations for doing so are particularly evident in the United States, where the cost of textbooks accounts for 26% of a 4-year degree programme[4] creating a strong economic argument for their adoption. Projects such as OpenStax have targeted subject areas with large national student populations, for example 'Introduction to Psychology', 'Concepts of Biology', 'Introduction to Sociology', etc. The books are co-authored and authors are paid a fee to author the books, which are peer-reviewed. The books are released under a CC-BY license, and educators are encouraged to modify the textbooks to suit their own needs. In terms of adoption the OpenStax textbooks had been downloaded over 120,000 times and 200 institutions had decided to formally adopt OpenStax materials, leading to an estimated saving of over US$30 million in a little over two years.[5] However, open textbooks are only one instantiation of the OER approach, and projects such as the Open University's OpenLearn focus on releasing more online, distance education focused content.

In summary then the OER movement has managed to grow substantially over the past fifteen years. It has released a vast amount of educational material, and established a range of implementation projects across the globe. In this time we have witnessed different phases, from startup, to growth and sustainability. This has happened in parallel with a number of related developments in the open education movement, namely the success of open access publishing, particularly through national mandates,[6] and the more recent popular attention garnered by so-called massive open online courses (MOOCs). This has created a context in which the OER movement views the next phase as one of becoming mainstream in educational practice. For example, the Hewlett Foundation White Paper (2013) on OER states that its goal is 'to pave the way towards mainstream adoption of OER in a manner that promotes greater, sustainable educational capacity,' and the theme of the 2015 OER conference in the UK was 'mainstreaming open education'.[7]

Research in OER is seen as a key component in facilitating this mainstreaming, as it provides evidence for benefits. OER Research has been increasing as the field matures. The OER Knowledge Cloud is a project gathering together all publications that relate to OER (https://oerknowledgecloud.org/), and an examination of the number of publications per year is shown in Table 1.

The data in Table 1 demonstrate a considerable increase in publications in 2007, as the initial projects became established and follow-on ones were implemented. This can be seen as the first expansion of the OER movement from a few key projects. Then in 2010 there is another sharp increase in the number of publications, which can be seen as a further expansion and acceptance of the OER approach.

Much of the early period of the OER movement was characterised by a lack of rigorous research however. The emphasis was on establishing OER projects, and developing content. For instance, a content analysis performed by the authors of the 2007 publications in the OER Knowledge Cloud revealed the categories and number of articles shown in Table 2.

This demonstrates that as the field began to grow, resources were focused on developing the projects and infrastructure required, along with theorising about the application of OERs. The sort of impact data that is desired by decision makers was often not resourced, or performed with a lack of independence and objectivism.

Year	No publications	Year	No Publications
2001	3	2008	58
2002	1	2009	67
2003	0	2010	153
2004	3	2011	121
2005	7	2012	167
2006	9	2013	205
2007	26	2014	183

Table 1: Number of OER publications per year, as represented in the OER Knowledge Cloud.

Category	No Publications
Project case study/announcement	6
Technical infrastructure	6
OER general discussion, guidelines	11
Research with impact data	3

Table 2: Types of OER publications in 2007.

Beliefs regarding the benefits of OER were often stated in publications, including their ability to:

• Radically reduce costs.
• Deliver greater learning efficiency.
• Promote continuous improvement of instruction and personalized learning.
• Encourage translation and localization of content.
• Offer equal access to knowledge for all.

However, empirical data to evidence these beliefs was usually absent. This was the impetus for the founding of the OER Research Hub.

OER Research Hub

The OER Research Hub is a project at the UK Open University which was funded by the Hewlett Foundation in 2012, to address this perceived need to develop a more robust evidence base for the impact of OERs. Drawing on previous research and in dialogue with the Hewlett Foundation, the project developed eleven hypotheses which represented some commonly stated beliefs and motivations regarding OERs. These were derived from previous experience, consultation with Hewlett Foundation and stakeholders, and analysis of common claims in OER literature. The full set of hypotheses is:

A – Performance: Use of OER leads to improvement in student performance and satisfaction.
B – Openness: The Open Aspect of OER creates different usage and adoption patterns than other online resources.
C – Access: Open Education models lead to more equitable access to education, serving a broader base of learners than traditional education.
D – Retention: Use of OER is an effective method for improving retention for at-risk students.
E – Reflection: Use of OER leads to critical reflection by educators, with evidence of improvement in their practice.

F – Finance: OER adoption at an institutional level leads to financial benefits for students and/or institutions.

G – Indicators: Informal learners use a variety of indicators when selecting OER.

H – Support: Informal learners adopt a variety of techniques to compensate for the lack of formal support, which can be supported in open courses.

I – Transition: Open education acts as a bridge to formal education, and is complementary, not competitive, with it.

J – Policy: Participation in OER pilots and programs leads to policy change at an institutional level.

K – Assessment: Informal means of assessment are motivators to learning with OER.

The project adopted a mixed methods approach. As well as gathering existing evidence onto an impact map (oermap.org), the project worked with 15 different collaborations, across four sectors: K- 12 (kindergarten through secondary school), Community College, Higher Education, and Informal Learning. Interviews, case studies, and quantitative data were gathered, but this chapter mainly reports on responses to surveys. A set of survey questions was created, addressing the 11 hypotheses. Although slight variations were permitted depending on context, the same pool of questions was used across a wide range of respondents. These included students in formal education, informal learners, educators at K12, Community College and Higher Education level and librarians. In total 21 surveys were conducted, with nearly 7,500 responses.

Key findings

This section will provide an overview of some of the key findings, according to three types of OER users: informal learners, formal learners and educators.

Formal learners

Formal learners in this instance are categorized as anyone who indicated they were enrolled in a formal offering with an education institution. The key factors influencing formal learners decision to use OER were all related to cost and ease of access, with 88.1% of respondents stating it was the opportunity to study at reduced or no cost that was a key factor, followed by the material being available any time (79.6%) and online (79.3%).

The survey revealed that many students (30% of those we surveyed) had studied their current formal subject in OER prior to signing up for a course and 52.7% supplemented their current study by using OER from another institution. There was broad coverage across all disciplines, with science being the most popular subject studied via OER (43.4%). Similarly there was a broad use of different types of OER, with video (79.2%) and open textbooks (79.1%) being the most popular formats.

The reported impact was often around attitudes to study rather than performance, with 61.9% stating that use of OER increased their interest in their subject, 60.7% increased satisfaction with the learning experience and 60.4% increased enthusiasm for future study. This is compared with only 38.9% who felt that it had improved their grades. When selecting OERs, the key indicators in choice were relevance to the students' particular needs or interests (72.6%), clear learning objectives (65.5%), and the reputation of the provider (60.6%).

Users were satisfied with their experience of OERs, with 83.5% stating they would study again with OER and 80% that they would recommend them.

Informal learners

In this context informal learners are defined as a learner who is not enrolled in a formal course of study. In reality, formal learners in one discipline can be informal learners in another, but this becomes a problematic interpretation, and so the clearer distinction was used in this work. For informal learners similar reasons for deciding to use OER were given, although the desire to study at no cost was more significant here with 89% of respondents stating this as a key factor. As with formal learners there was a broad coverage of topics, but the dominant disciplines were computer science (31.8%) and economics (30.6%). The same three elements guided choice as for formal learners, namely relevance, learning objectives and reputation. Here, OER can be perceived as plugging a specific gap in knowledge or skills. Less important were reviews of OER or personal recommendations. Open or Creative Commons licensing allowing adaptation was only thought important by a minority of 13.9%. Informal learners used a wide range of support methods, such as notetaking, discussion in social media and blogging, although not more than the formal learners.

As with formal learners, informal learners were likely to study using OER again (78.5%) and recommend them to others (80.4%). Only 24.6% stated that studying OER would make them more likely to take a paid for course, but this could still represent a very effective recruitment method, compared with other forms of marketing. Given the potential large numbers of OER users, this could translate into substantial numbers of new students for universities. However, this should be countered with the 19% who stated that studying OER had made them less likely to take a paid for course, either because the OER were sufficient to meet their learning requirements, or they decided study was not something they wished to pursue. These future study patterns for formal and informal learners are summarized in Figure 1.

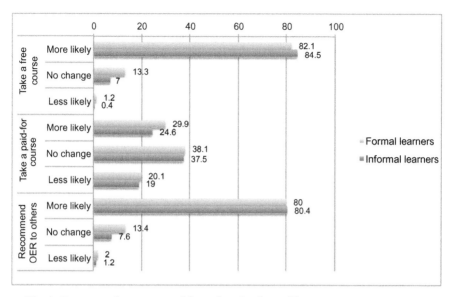

Fig. 1: Future study patterns of formal and informal learners.

Educators

When asked about the impact of OER on their teaching, there was strong agreement from both full and part-time educators working at a range of different institutions that OER gave them a broader range of teaching methods (64.3%) and caused them to reflect on their practice (59.4%). Corresponding to the views of learners themselves, educators believed that use of OER increased learners' satisfaction with the subject (62.1%), increased their interest (60.8%), and increased experimentation with new ways of learning (60.3%).

By far the most prominent motivation for using OER was to get new ideas and inspiration for teaching (78.2%). This is compared with only 40.7% who stated that their motivation was to get assets for use in the classroom, which is often presumed to be the main driver for OER.

The same three factors were deemed important for educators as for learners in choosing a resource, namely relevance, learning outcomes, and reputation. Only 43% stated that having an open license was significant. The open licenses of OER are seen as a key differentiator between them and general online resources, as it allows users to reuse and adapt. Whilst the presence of such a license may not be that significant to learners, it might have been presumed to be of significance to educators, but it was ranked 12th out of 16 factors in importance.

Most educators believed that use of OER saved their students money (73.1%), although this was lower for formal learners themselves (60.9%). This may arise because some formal learners purchased the physical textbook also.

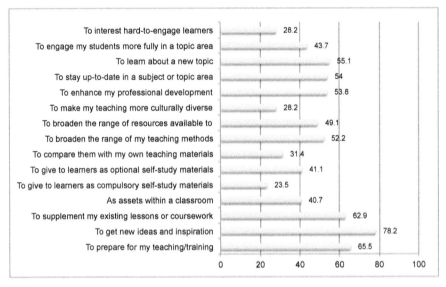

Fig. 2: For what purpose do educators use OER?

For all users, the biggest barrier to use of OER was finding suitable resources, and OER repositories were used very little compared with resources such as YouTube, Khan Academy, and TED talks. Another finding that was consistent across all groups was the comparatively high level of adaptation. This might be expected to be high amongst educators (79.8%), but was also found in formal learners (77.3%), and informal learners (84.7%). What constitutes adaptation varies for these users, and is an area that requires further investigation. This is in contrast to other research which found previously low levels of adaptation.[8] For some users adaptation means using the resources as inspiration for creating their own material, as this quote illustrates:

> 'What I do is I look at a lot of free resources but I don't usually give them directly to my students because I usually don't like them as much as something I would create, so what I do is I get a lot of ideas.'

Math Teacher, Grade 11

For other users, adaptation is more direct, editing or reversioning the original, or aggregating elements from different sources to create a more relevant one, as this quote demonstrates:

> 'The problem where I teach now is that we have no money; my textbooks, my Science textbooks are 20 years old, they're so out-dated, they don't relate to kids (…) so I pick and pull from a lot of different places to base my units; they're all based on the Common Core; for me to get my kids to

meet the standards that are now being asked of them, I have no choice, I have to have like recent material and stuff they can use that'll help them when they get assessed on the standardised test.'

Math & Science Teacher, Grades 7–8

And for others, adaptation may be taking an existing resource and placing it in a different context within their own material, for example:

'I will maybe look and find an instructional video that's maybe 2 or 3 minutes long that gets to the point better than I could, and I would use it, or I will look for lessons and if they are for Grade 5 or Grade 3 I don't use all of it, I just adapt it, I take out what I don't want and rearrange it.'

Math teacher, Grade 2

What this suggests is that one impact of openness is that it allows a continuum of adaptation to develop, ranging from adapting ideas for teachers' own material to full reversioning of content.

Discussion

A general picture of OER usage emerges from this research. There was a high degree of satisfaction with OER across all types of user, with a large percentage willing to access further OER and to recommend them. However, OER brand recognition was weak compared with other popular resource sites, and finding appropriate OER was a major obstacle. Use of OER increases satisfaction and engagement with learning and is seen as saving students money. Users look for relevance, reputation and clear learning outcomes when selecting OER. The use of OER is not confined to one or two disciplines, with all subjects well represented, and a range of formats are accessed, although video remains the most significant.

These findings themselves would be significant and useful to the OER community. But what this research also reveals is a more nuanced, subtle picture of OER usage than had often been supposed. Many of the benefits of OER being portrayed in this usage are not the primary ones of improved cost savings or improved performance. While these are important gains, and ones that will have strong leverage on policy, the benefits discovered by the OER Research Hub may have more long lasting effects. For example, the use by students in formal education to both trial and then supplement their learning may have an impact on student retention. Combined with a quarter of informal learners stating that they were more likely to study a paid for course after using OER, and the sustainability model for OER may be one that can be couched in terms of recruitment and retention (compared with the sale of additional services as seen with MOOCs).

Another factor that is important to students, but also increasingly used as a metric to rank universities, is student satisfaction. The impact of OER on emotive aspects related to learning such as satisfaction, enthusiasm, and confidence could be of greater relevance than cost savings.

A similarly under-reported benefit for educators is the manner in which OER cause them to reflect on their own practice, and to broaden their teaching approaches. The use of OER is often couched in terms of benefit to the learner, but the impact on educators could be equally significant. This needs to be balanced with the relative unimportance users placed on open licenses. An open license potentially allows an educator to take existing content from several different sources (for example, different open textbooks, videos) and adapt this to their own context, to produce something that is ideally suited to their specific learning outcomes in a way that a generic textbook can never be. However, if awareness of open licensing remains relatively low, these pedagogic possibilities of open content will not be realized.

What this research highlights beyond the use of OER, is the significance of research itself in an emerging discipline. Open education in its current guise is still in its infancy as a discipline, and research plays an important role in how it evolves. In the early stages researchers are often in the role of advocates, but as the field matures, more objective research can be undertaken. It is necessary for a field to gain momentum for it to commence from a set of beliefs and assumptions about the potential impact. These can later be tested as the OER Research Hub and other groups such as the Open Education Research Group[9], and ROER4D[10] have done. Just as important however is to research into unexpected usage. The type of actual use of OER that this research found helps inform the sector. For an emerging discipline seeking to become part of mainstream this is important as it helps inform strategy, policy, and direction in a more direct manner than is possible with more established domains.

In order to realize the ambition of mainstreaming OER then there are two research related aspects. The first is that this type of objective, impact research becomes part of all implementation projects. The second is that it is communicated effectively to help shape strategy. In addition, a number of specific recommendations might be drawn from this work. We believe that increasing the 'brand' of OER through joint initiatives, sequenced activities, and improved marketing could significantly improve uptake and adoption. Making the business case for OER beyond the immediate cost savings of textbooks is also important, for instance emphasizing the manner in which all students benefit by trialing subjects and complementing formal study, and educators use OER for development.

Notes

[1] Weller, 2014.
[2] Wiki, n.d.
[3] Wiley, 2011.
[4] GAO, 2005.

5   OpenStax College, 2014.
6   SPARC, 2015.
7   OER conference, 2015.
8   Wiley, 2009.
9   Open Education Research Group, n.d.
10   ROER4D, n.d.

References

Government Accounts Office. (2005). *College textbooks: Enhanced offerings appear to drive recent price increases.* Retrieved from: http://www.gao.gov/assets/250/247332.pdf

William and Flora Hewlett Foundation. (2013). *White paper: Open educational resources breaking the lockbox on education.* Retrieved from: http://www.hewlett.org/library/hewlett-foundation-publication/white-paper-open-educational-resources

OER Conference. (2015). *Mainstreaming open education.* Retrieved from: https://oer15.oerconf.org/

Open Education Research Group. (n.d.). Available at http://openedgroup.org/

OpenStax College. (2014). *Our textbooks have saved students $30 million.* Retrieved from: https://www.openstaxcollege.org/news/our-textbooks-have-saved-students-30-million

ROER4D. (n.d.). Available at http://roer4d.org/

SPARC. (2015). *Analysis of funder open access policies around the world.* Retrieved from: http://sparceurope.org/analysis-of-funder-open-access-policies-around-the-world/

Weller, M. (2014). *Battle for open: How openness won and why it doesn't feel like Victory.* London: Ubiquity Press. DOI: http://dx.doi.org/10.5334/bam

Wiki. (n.d.). Available at https://wiki.creativecommons.org/wiki/OER_Policy_Registry

Wiley, D. (2009). Dark matter, dark reuse, and the irrational zeal of a believer. [Web log comment] Retrieved from: http://opencontent.org/blog/archives/905http://opencontent.org/blog/archives/905

Wiley D. (2011, August 26). The $5 textbook. [Web log comment]. Retrieved from: http://utahopentextbooks.org/2011/08/26/the-5-textbook/

Are OE Resources High Quality?

Regan A. R. Gurung

University of Wisconsin-Green Bay, gurungr@uwgb.edu

Editors' Commentary

Open Education is a relatively new phenomenon. As such, it will have skeptics and detractors. Among the common doubts concerning OERs are questions related to their quality. Indeed, in many markets price is a partial indicator of quality. The free-of-charge nature of open education means that prospective consumers have one less indicator of potential quality with which to judge these resources. It is in this context that author Regan Gurung tackles the foundational question of whether OERs are any good. He points to preliminary research on the topic. While early research shows mixed results Gurung also offers insights as to how we ought to be evaluating quality more broadly. Among his recommendations are outcomes research, materials that have been peer reviewed, and expert authorship.

How do you know when you have a quality product? If it is a meal at a fine restaurant or a beverage concocted by a skilled mixologist there are some dead giveaways. It is expensive. It tastes good. It looks good. It makes you feel good. Can we use the same rubrics for Open Educational Resources (OERs)? How do you know if an OER is high quality? There are some direct similarities to a good meal but of course, many more differences. As someone relatively new to the whole concept and use of OERs, this seemingly simple question provoked some deep digging. I have some answers and I'd like to share some factors for your consideration.

There are many ways to learn. I like to think that armed with a curious mind and the right resources and motivation, anyone can learn by themselves. Of

---

How to cite this book chapter:
Gurung, R A R. 2017. Are OE Resources High Quality?. In: Jhangiani, R S and Biswas-Diener, R. (eds.) *Open: The Philosophy and Practices that are Revolutionizing Education and Science.* Pp. 79–86. London: Ubiquity Press. DOI: https://doi.org/10.5334/bbc.f. License: CC-BY 4.0

course, when we think of learning we do not think of the solo pursuits of motivated individuals. We tend to think of schools and colleges. We all want a quality education for our children and lists extolling the virtues of select institutions to deliver such educations abound (think US News and World reports). When we dive deeper into what makes college a quality experience we easily settle on faculty and the classes they teach. While master teachers can inspire with their passion and masterfully deliver content, most students rely heavily on course materials the faculty assign (though the students may not always read all of it) to solidify content acquisition. Sure classroom discussion and deep processing may inspire the master student, but it is the textbook that is the crutch of the average student. Consequently, the quality of course material is of tantamount importance. Yes, of course you know the quality of the material is important, but I state this at the outset because I fear that unconsciously at least, some instructors may believe their brilliance transcends the need for quality course materials. Passionate, organized, motivating, and knowledgeable instructors who build student-rapport are important to learning, but quality material is important as well. I am getting off the soapbox now.

Is it Expensive?

Once upon a time, you could rely on the simple heuristic that pricey equals quality. As the social psychologist Robert Cialdini (1993) showed some time ago, we humans are easily influenced by price. He tells of a sales trick where cheap dollar jewelry actually sold more when its price was hiked up many times. People actually paid more for a pair of earrings when the price tag was changed to be higher. The expensive meal must be better than the cheap meal. The gourmet taco for ten dollars should be better than the two-dollar food truck taco. This is sometimes the case and reasons are clear. Gourmet taco making chefs may be trained to best coax flavors from hitherto untried pairings. They may use more expensive ingredients. A similar intuition accompanies the common belief that expensive textbooks are better. After all, expensive textbooks have the backing of major publishing companies who have invested large sums of money to ensure quality products right? The development editors, slew of peer reviewers examining every draft of every chapter, and focus groups should ensure a quality product right? There is some evidence that this is the case.

There is no doubt that most expensive books come with a lot of bells and whistles. Big Publisher Books (BPBs) as I like to call them, are multicolored affairs. They are packed with pictures, can often afford the rights to comics and cartoons, and come with a wide array of textbook technology supplements (online quizzes, etc.). BPBs also tend to have well known authors, recognizable for their research chops, and undergo a long arduous process of review. We assume that for all of these reasons, the expensive BPB must be high quality. We can, and I have, tested many of these assumptions.

Do students like all BPBs the same?

Do students like some books more than others? This question is difficult to test without comparisons across class sections or universities, as each class uses only a single book. The good news is a number of such tests exist.

One of the best way to compare books is to have the same student read a number of books. I did a study some years ago where I had students rate a number of most adopted textbooks in the introductory psychology market.[1] I had students first read a chapter from each of seven textbooks (yes, it was a long study and I paid each student US$10.00), and then rate the book using a 28 item survey. I randomized the order of the books so students saw different books at different points in the line up. The bulk of the questions came from the validated *Textbook Assessment and Usage Scale*[2] and measured opinions about the figures, tables, photographs, research examples, application examples, pedagogical aids, visual appeal, and writing quality. Students did differentiate between texts rating some books better than others. From a student opinion perspective, not all BPBs are rated the same.

In a number of national studies, colleagues and I had students rate the quality and helpfulness of their introductory psychology textbooks.[3] The students also rated how much they thought they learned from the books and then took a quiz on material taken from the learning and biological psychology chapters. All students took our quiz so we had a common measure of learning. All students in our studies used BPBs. We found some significant differences in how students rated how much they learned and their textbooks (some books were rated higher in quality and helpfulness than others), but quiz scores were similar. Regardless of which textbooks the student was using, their quiz scores on our test did not vary. In this case, students did like some books better but did not learn better from some books. Given learning should be an instructor's key focus, this outcome bears a great focus.

How does Learning Vary Between Books?

Fine you may say, so students like some books better than other. Regardless of preference do students learn better from some books over others? I took the lab study mentioned above a step further. In a second study I had students come in to my laboratory, read chapters from two different books, and take a quiz on what they had read. Students read a biology chapter from the first book and the learning chapter from the other book. They also rated two additional books. When I compared the quiz scores I did not find any significant differences between quiz scores regardless of the chapter tested or textbook used. Students seem to learn the same in a lab test of textbooks but of course the lab is not the classroom. Wouldn't it be great to be able to control for instructor and lecture content but still test at least two books?

One semester some years ago I did exactly this. I selected two well adopted BPBs that varied in 'look.' One was in what is called a magazine format: lots of pictures, a layout resembling *Cosmopolitan* or some other glossy. The other was a standard brief edition of an introductory psychology book. Both books had a similar reading and difficulty level. I worked with my campus bookstore and gave students in my Intro psychology class a choice. They could buy whichever book they wanted. The bookstore even set up a little booth where they could page through each book and decide which one they wanted. About 60% of the class picked the magazine format book. When it came to exam time, I wrote two forms of the exam tailoring exams to the book bought by the student – the bulk of the questions were the same but a small number explicitly mentioned material from the book the student picked. At the end of the semester, the students rated the magazine format book higher in visual quality (no surprise) but there were no differences in exam scores. Again, students do not seem to learn differently from different BPBs. But now for the big question: How do OERs compare?

OERs and BPBs: Head to Head

Research testing OERs and BPBs is still in its infancy but a number of studies have attempted to assess if OER use influences student learning.[4] The story is mixed. The best studies using standardized or similar exams[5] show no differences in exam scores between OER users and BPB users. A number of studies demonstrated questionable statistical validity. For example, although one suggests OER users show higher exam scores in psychology classes, the study did not include statistical tests of the difference and the exams used (across two different semesters) may not have been equivalent in difficulty.[6] In another study of OER use in English classes, authors state students who used OERs scored higher on reading tests than peers using traditional means, but again no statistical tests were reported and it was unclear if the assessments were comparable.[7] In short, the current research comparing OERs and BPBs is fraught with limitations and validity issues.

In an attempt to transcend the limitations of extant studies, I recently compared a group of OER users to BPB users. In collaboration with the NOBA project, instructors at seven different schools invited their students to take part in a study of learning. Over a thousand students took part in the study. A little over half of the students used an OER (a NOBA intro psych textbook) and the rest used a BPB. I compared student perceptions of the material and similar to the other studies discussed above, also had all students take the same test (mine). In one of the first studies of its kind pitting OER against BPBs, students using an OER rated the material as more applicable to their lives. Score one for OER. Students using the OER also rated the quality of the photographs, figures, tables, and boxed information as lower than the BPB. Score one for

BPB. When it came to study aids, writing quality, the examples, research studies or the extent to which components of the textbook helped students understand the material, there were no differences in quality between OER and BPBs. Given that the OER was free and the BPBs on average cost over US$100, multiple major scores for OER here. This finding was tempered by test score data. Students using a BPB did significantly better on the quiz. Unfortunately, this finding is potentially contaminated by the reality of the study test items coming from the psychology Advanced Placement exam, something that most BPB testbanks also draw from. A better test of OER users to BPB users would come from using a more neutral set of test questions (a study currently in progress).

It is important to note that the playing field is not equal. There are some significant differences in the current state of many OERs and BPBs. Whereas these differences may not be as important for upper level classes, the differences may be particularly important for lower level classes. In general, BPBs do tend to have higher production value. In addition to more color and comics mentioned before, they also have numerous pedagogical aids built into them. Not only do the books themselves have many features such as multicolored fonts, boxes, and running glossaries, but the books also have numerous textbook technology supplements. These study aids on the publisher websites allow students to test themselves on the material, often in a variety of engaging formats such as crosswords, matching games, and application questions. These textbook supplements do help students learn.[8] Many publishers also provide ready-made flashcards to accompany the book, something few if any OERs provide (OERs are quickly catching up though). Whereas these pedagogical aids may not necessarily always aid learning,[9] the perception that they do is often enough to get instructors to adopt BPBs and rationalize the price tag of the same. Finally, whereas many BPBs are also available online students tend to still prefer hard copy books. Most OERs are online by default and whereas some OERs afford a hardcopy version (for a small charge), few of the hard copy paper versions are as colorful as even the cheapest BPBs. While one should not judge a book by its cover, it is possible that students associate flashy production values with quality and are more likely to read a book awash in multiple cartoons, graphics, photographs, and figures.

So the story so far is this. There are some differences between the major BPBs in terms of how students' perceive them though no significant testing differences. There is a dearth of research comparing OER learning/test scores with learning/test scores on BPBs (most studies assess cost savings and perceptions),[10] though one study suggests BPB users may have an advantage on standardized tests. In short, there is currently no definitive answer to whether the quality of OERs especially as compared to BPBs and as measured in learning is high. What then can we use to measure quality? It is time to return to the usual suspects.

One Open Educational Resources site nicely summarizes the main places to look.[11] When assessing the quality of an OER, you can look at the reputation of the author of the piece, the institution or body curating the piece, the accuracy, and also factors such as accessibility and fit for purposes. The last two factors may not directly influence learning, one of my main criteria for 'quality' but are important nonetheless. When we talk about quality in higher education we tend to rely on the credibility of authors and the peer review process and this is where I put my money. If you want a quality OER, the fact is that it is going to be difficult to get learning outcome data. In fact, there is little learning outcome data for the use of BPBs but faculty still adopt these books. One of the major reasons why many faculty do adopt BPB books is because other faculty do (crazy circular logic on one hand). If many people adopt BPBs they *must* be good is the thinking. If it is put out by a publisher whose name is recognizable it *must* be good. If it written by an author who is familiar, it *must* be good. In fact, these are all empirical questions that are never really tested. The market research that big publishers cite and the student and faculty endorsements peppering the back covers and promotional materials of BPBs rarely (if ever) represent true comparisons of learning. To be fair, true comparisons of learning are difficult. A variety of factors – the student, the teacher, the textbook- all influence learning, which makes such research difficult.

So where does that leave me? If I know the author of an OER has a strong reputation and I know the piece has been reviewed by peers that will make me more likely to entertain the use of the OER. Fortunately openly published reviews by faculty colleagues is becoming more common (e.g., Open Textbook Network). Beyond that, we faculty have the responsibility to monitor the accuracy of material, something difficult to do when you are using resources for a class whose breadth expands beyond your own personal expertise (e.g., Introductory classes). The more faculty who use OER, the more these materials will be vetted. One of the most appealing features of OERs is that users are invited to modify the resources and with more faculty using OERs the better these resources will be. In many ways this is a form of academic freedom that will preclude having instructors bending the course to conform to a textbook. The sifting and winnowing of material will help OERs evolve. BPBs have inherited a veneer of quality not based in empirical tests of their links to learning. Well curated OERs, those where the writing and content is monitored and reviewed by peers and contributed by credible sources, deserve to likewise bask in the reflected glory of BPBs. My in-depth perusal of many OERs in psychology and research on faculty perceptions of OERs[12] show that the OERs are ready for their time in the spotlight while scholars of teaching and learning work to assess true quality of all educational resources. OERs are tasty meals at the right price, free.

Notes

[1] Gurung & Landrum, 2012.
[2] Gurung & Martin, 2011.

[3] Gurung, Daniel & Landrum, 2012; Gurung, Landrum & Daniel, 2012.
[4] Allen, Guzman-Alvarez, Molinaro & Larsen, 2015; Bowen, Chingos, Lack & Nygren, 2014; Fischer, Hilton, Jared-Robinson & Wiley, 2015; Hilton, Gaudet, Clark, Robinson & Wiley, 2013; Hilton & Laman, 2012.
[5] Allen et al., 2015; Bowen et al., 2014; Hilton et al., 2013.
[6] Hilton & Laman, 2012.
[7] Pawlyshyn, Braddlee, Casper & Miller, 2013.
[8] Gurung, 2015.
[9] Gurung, 2004.
[10] Open Education Group Publications, n.d.
[11] Open Educational Resources, n.d.
[12] Allen & Seaman, 2014.

References

Allen, G., Guzman-Alvarez, A., Molinaro, M., & Larsen, D. (2015). Assessing the impact and efficacy of the Open-Access ChemWiki Textbook Project. *Educause Learning Initiative Brief.* Retrieved from https://net.educause.edu/ir/library/pdf/elib1501.pdf

Allen, I. E., & Seaman, J. (2014). *Opening the curriculum: Open educational resources in U.S. higher education, 2014.* Retrieved from http://www.online learningsurvey.com/oer.html.

Bowen, W. G., Chingos, M. M., Lack, K. A., & Nygren, T. I. (2014). Interactive Learning Online at Public Universities: Evidence from a Six-Campus Randomized Trial. *Journal Of Policy Analysis And Management, 33(1),* 94–111.

Cialdini, R. B. (1993). *Influence: The psychology of persuasion.* New York, NY: Morrow.

Fischer, L., Hilton, J. I., Robinson, T. J., & Wiley, D. A. (2015). A multi-institutional study of the impact of Open Textbook adoption on the learning outcomes of post-secondary students. *Journal Of Computing In Higher Education, 27(3),* 159–172.

Gurung, R. A. R. (2004). Pedagogical aids: Learning enhancers or dangerous detours? Teaching *of Psychology, 31,* 164–166.

Gurung, R. A. R. (2015). Three investigations of the utility of textbook teaching supplements. *Psychology of Learning and Teaching, 1,* 48–59.

Gurung, R. A. R., Daniel, D.B., & Landrum, R. E. (2012). A multi-site study of learning: A focus on metacognition and study behaviors. *Teaching of Psychology, 39,* 170–175. DOI: https://doi.org/10.1177/0098628312450428

Gurung, R. A. R., & Landrum, R. E. (2012). Comparing student perceptions of textbooks: Does liking influence learning? *International Journal of Teaching and Learning in Higher Education, 24,* 144–150.

Gurung, R. A. R., Landrum, R. E., & Daniel, D. B. (2012). Textbook use and learning: A North American perspective. *Psychology of Learning and Teaching, 11,* 87–98.

Gurung, R. A. R., & Martin, R. (2011). Predicting textbook reading: The textbook assessment and usage scale. Teaching of Psychology, 38, 22–28. DOI: https://doi.org/10.1177/0098628310390913

Hilton, J. I., Gaudet, D., Clark, P., Robinson, J., & Wiley., D. (2013). The adoption of Open Educational Resources by one community college math department. *International Review Of Research In Open And Distance Learning, 14*(4), 37–50.

Hilton, J., & Laman, C. (2012). One college's use of an open psychology textbook. *Open Learning, 27*(3), 265–272.

Open Education Group Publications. (n.d.). Available at http://openedgroup.org/publications

Open Educational Resources. (n.d.). Available at https://openeducationalresources.pbworks.com/w/page/24838164/Quality%20considerations

Pawlyshyn, N., Braddlee, D., Casper, L., & Miller, H. (2013). *Adopting OER: A case study of cross-institutional collaboration and innovation.* Retrieved from http://er.educause.edu/articles/2013/11/adopting-oer-a-case-study-of-crossinstitutional-collaboration-and-innovation

Open Practices

Opening Science

Brian A. Nosek

University of Virginia & Center for Open Science, nosek@virginia.edu

Editors' Commentary

When many people think about 'open' their minds jump immediately to notions of free resources, or perhaps to the ability to collectively edit resources, as in the case of Wikipedia entries. Often overlooked in discussions about open are applications to science. In an era where science is plagued by problems such as non-replication, p-hacking, and data fabrication open offers a solution for scientific self-correction. In this chapter, author Brian Nosek showcases a number of ways in which scholarly collaborations and professional societies are using open to improve the quality of research. He focuses especially on evaluating empirical evidence, offering better training, and providing simple incentives for increasing the openness, transparency, and rigor of science.

When my 9 year-old daughter Haven learned about the scientific method in school, she learned that a scientist starts by observing what happens in the world. After collecting enough observations, the scientist generates a question and perhaps a prediction. Then, the scientist designs a study to investigate the question and test the prediction. After collecting data, the scientist learns whether the results are consistent with the prediction or not. Either way, the scientist learns something. The scientist shares the study and data with others so that they can learn too, or try the study themselves to see if they get similar results. Finally, the data are observations for making new questions and predictions.

At dinner, Haven described how her class tried out the scientific method themselves by guessing how many times a coin will show heads when they each

How to cite this book chapter:
Nosek, B A. 2017. Opening Science. In: Jhangiani, R S and Biswas-Diener, R. (eds.)
    Open: The Philosophy and Practices that are Revolutionizing Education and Science.
    Pp. 89–99. London: Ubiquity Press. DOI: https://doi.org/10.5334/bbc.g. License:
    CC-BY 4.0

flipped it ten times. They observed, predicted, measured, evaluated, shared what they did with the class, collated the observations, and began again. It wasn't hard, it made sense, and it was completely exciting.

When I said that a lot of science doesn't actually work that way, Haven was puzzled. What is different? Well, sometimes I don't share the results of studies we do with others. Why not? Well, they don't come out the way we expected. But, you still learned something didn't you? Well, yes, but other scientists are not as interested in those studies. But, isn't it important for other people to know about when you were wrong so they can learn or try it out themselves? Sure, but it takes a lot of time to share what we did. But, if you aren't going to share it, then why did you do it? I was running out of answers, and Haven was losing interest.

The ideals of science versus the reality of science

How science is supposed to work and how it actually works are not the same thing. In a survey of more than 3,000 practicing scientists, more than 90% endorsed the norms of science such as transparency, skepticism, and disinterestedness over the counternorms such as secrecy, dogmatism, and self-interestedness.[1] When asked how they behaved on a daily basis, fewer but still most scientists said that they behaved more according to the norms than the counternorms. But, when asked how others in their discipline behaved, most respondents perceived their peers to behave according to the counternorms over the norms.

This cynicism about science among scientists is not unfounded. A substantial body of evidence shows that transparency and sharing of data and methodology is the exception rather than the rule,[2] and that a variety of suboptimal research practices undermine the integrity and reproducibility of the published literature. Sub-optimal practices include underpowered research designs, selecting reporting of results, and failing to distinguish between exploratory and confirmatory approaches.[3]

At the same time, the survey revealed that scientists want to behave according to the norms of science. The problem is that the culture of science has skewed the incentives such that researchers perceive that they are not rewarded for transparent, reproducible research, but rather for shaping data – even at the cost of accuracy – to make the most exciting, bold claims possible in order to achieve the reward of publication.[4]

Origins of the Center for Open Science

It is uncomfortable to be in a culture that is perceived to be misaligned with one's values. How can we change that? One step is to show that the norms and values are actually shared, even if they are not rewarded in practice. A second step is to make it easy for people to behave according to their values, and particularly to not be punished for doing so. A third step is to surface when people are practicing the valued behaviors to signal to others that it is possible, practical, even prevalent. A fourth step is to show that the counternorms are having negative consequences on the quality of research, providing a means of reinforcing the normative behaviors. And, a final step is to shift the cultural incentives so that they actually support and reinforce the normative behaviors.

Since I joined the faculty at the University of Virginia in 2002, the members of my laboratory have contributed to addressing the first four steps, if only for ourselves. We talk about our scientific values and the practical and cultural barriers to practicing them. We look for ways to address the barriers with existing tools. And, when nothing exists to support the ideals we aim to practice, we create new tools. And, while, the members of my laboratory have no control over the cultural incentives, we also do not accept the status quo. To maintain integrity with our values, we adopted an ethic of pragmatic idealism. What does that mean? Each day, we ask ourselves 'what can we do today to behave more closely to our ideals while still working and succeeding in the present culture?'

For example, low-powered research is a pervasive problem and increases the likelihood of both false negatives and false positives.[5] That is a lose-lose situation. No good that comes from collecting smaller samples than needed to properly evaluate our research questions. To ensure that we conduct properly powered research, we created and maintain a website called Project Implicit to collect data via the internet.[6] With an engaging blend of education and research, the site attracts about 1 million participants per year. That was a couple orders of magnitude more successful than we could have anticipated, but it solved our power problem for the kinds of research that could be administered via the web.

In 2006 and 2007, we wrote grants to tackle another gap between our values and practices – transparency. We wanted to improve sharing our research workflow, materials, and data, but the infrastructure didn't exist to support this practically. Every option we found added a substantial amount of work, and we were already busy enough. We wanted to develop tools so that sharing was a benefit, not a burden. Unfortunately, our grant applications were not sufficiently compelling and they failed. Partly, we may not have had the right pitch, but another challenge might have been that the pitch was ill timed. One reviewer criticized the proposal with a simple point, 'Researchers don't like to share their data.' In 2007, that was a compelling argument. Ultimately, we shelved the idea, and did the best approximation for our sharing goals by using existing tools.[7]

The idea was re-energized in 2011 when Jeff Spies, a senior member of the laboratory, was choosing among a variety of possible dissertation topics. As a software developer prior to coming to graduate school in quantitative psychology, Jeff had a strong sensibility for building tools that could improve his and others' research workflow. Jeff recognized that building infrastructure was a very unusual choice for a dissertation topic in psychology, but simultaneously saw huge potential in building tools to make the research lifecycle more transparent. Jeff jumped in with both feet to build and evaluate the Open Science Framework[8] as his dissertation project.

At the same time, the lab was starting a study called the Reproducibility Project: Psychology to evaluate the growing concerns about reproducibility of scientific research.[9] For this project, we planned to conduct replications of a sample of published studies and compare our findings with the results of the original studies. However, we could not conduct enough replications on our own to get a meaningful sample of studies. So, we opened the project to the research community for anyone to join.

In our good fortune, there were many others with similar interests. Moreover, many were willing to donate some of their time to conducting replications. The project became a collaboration of 50, then 100, then 150 researchers. And, with the alpha version of the Open Science Framework released, we had infrastructure to coordinate the large collaboration and conduct the project transparently.

We didn't have grant funding, but I had resources to support these projects from giving presentations to organizations about implicit bias, my substantive area of research. Unexpectedly, in the Fall of 2012, the Reproducibility Project: Psychology and Open Science Framework kindled interest from funders that was not present just a few years earlier. Following some press coverage, we were contacted by multiple foundations. After a series of emails, virtual demos, and a visit to Houston, the Laura and John Arnold Foundation gave us a grant for US$5.25 million to launch a non-profit called the Center for Open Science (COS). There had been two self-funded lab projects, and suddenly there was a well-funded non-profit. This might seem a rather abrupt acceleration. Yes, yes it was.

We launched COS with a mission to increase openness, integrity, and reproducibility of scientific research. In its first three years of operation, COS received US$18 million from a combination of private and federal funders to support its mission. The Open Science Framework and Reproducibility Project: Psychology provided a foundation to which we added a full suite of investigations, products, and services to evaluate and improve research practices. COS provides free and open tools and services to the stakeholders in science – funders, journals and publishers, universities, societies, research producers, and research consumers.

COS's Approach

COS has three teams – metascience, community, and infrastructure – that form the basis of its strategy to increase openness, integrity, and reproducibility of scientific research.

Evidence

The metascience team conducts and supports scientific research about scientific practices. The goal is to accumulate evidence about the problems and opportunities to improve research practices, and to evaluate interventions aiming to address those problems. For example, the Reproducibility Project: Psychology estimated the reproducibility rate in a sample of psychology articles and explored predictors of reproducibility. The completed project involved 270 co-authors and included 100 replications.[10] That project produced a spin-off Reproducibility Project for cancer biology,[11] and related efforts conducting replications of the same research protocols in many labs to examine variability in replicability[12] and independent analysis of the same data by many teams to examine variability in analysis decisions and their impact on observed effects.[13] These investigations provided insights about the current state of research practices.

COS also empirically evaluates whether its initiatives have an impact on research practices, positive or negative. For example, a recent grant from the National Science Foundation is supporting a randomized trial to evaluate the impact of receiving training to use the Open Science Framework. Research laboratories at University of California-Riverside will be randomly assigned to receive the training or not (and an orthogonal factor will evaluate training on the responsible conduct of research). Likewise, we evaluated the impact of one of our first and simplest initiatives to incentivize openness – offering badges on journal articles to signal open practices.[14] The journal *Psychological Science* adopted badges to signal open data, open materials, and preregistration on January 1, 2014. From 2012 to 2013, before badges, approximately 3% of the journal's published articles had open data. After introduction of badges, open data practices increased each half year starting in 2014 reaching 38% of published articles in the first half of 2015. Comparison journals maintained very low data sharing rates across the entire time period. Direct evidence for the effectiveness of initiatives to improve openness will facilitate their adoption and impact.

Training

Researchers possess the values of transparency and reproducibility but if they do not have appropriate training, they may not be able to translate them into practice. The community team produces articles or chapters providing guidance on reproducible practices.[15] The team also creates and maintains free text and video content on the Open Science Framework for improving reproducibility and transparency, and conducts webinars and on-site trainings for laboratories, departments, and other research groups. Finally, the team offers free one-on-one virtual consulting to address statistical or methodological challenges related to reproducible research.

Incentives

Even the combination of values and training may not be sufficient for increasing openness and reproducibility in daily practice. Academic researchers are busy and have significant pressures to be productive and publish in order to be competitive for jobs, earn tenure, and advance in their careers. Without realigning the incentives shaping researchers' behavior, even the best interventions may not stick. The community team works with stakeholders in the scientific community – particularly funders and journals/publishers – to strengthen incentives for open, reproducible practices.

Badges to acknowledge open practices are a simple, low-risk, low-cost nudge toward openness.[16] Badges offer no onerous requirements of journal or authors, they just offer an opportunity for authors to signal that they met specifications for open data, open materials, or preregistration.[17] Badges also offer an opportunity for journals to signal that such practices are valued, even if they do not directly impact publication decisions.

Registered Reports shift publishing incentives more fundamentally to address publication and reporting biases.[18] For journals that adopt Registered Reports, authors can submit the introduction and methodology for peer review before the research is conducted. Peer reviewers evaluate the importance of the research question and the quality of the methodology that will investigate it. The journal provides in-principle acceptance to submissions that survive review.[19] *After that*, the researchers collect the data, analyze it, and report what they found. As long as they conduct the methodology effectively, the results are published whatever the outcome. In standard peer review, the incentives focus on having beautiful results, even at the cost of accuracy. In Registered Reports peer review, the incentives drive researchers to ask the most important questions and have the most beautiful methodology to evaluate those questions. Already more than 20 journals have adopted Registered Reports, and COS facilitated publication of a special issue of *Social Psychology* in 2014 demonstrating the viability of the approach.[20]

Another approach to shifting the incentives is to directly incorporate incentives for transparency and reproducibility into publication and funding. The Teaching of Psychology (TOP) Guidelines achieve this by defining eight modular transparency standards for journals and funders to adopt as policies for their authors and grantees.[21] The guidelines also ease the barrier to adoption by having multiple levels of stringency so that journals and funders choose to nudge toward openness, or make a bolder requirement of their authors and grantees, depending on the circumstances for their journal or discipline. As of early 2016, more than 50 organizations and 500 journals were signatories to the TOP Guidelines.[22]

One of the TOP Guidelines is preregistration of analysis plans to promote a clear distinction between confirmatory and exploratory research.[23] Preregistration is the law in clinical trials, but is relatively unknown in basic or preclinical sciences. Of the variety of new practices to promote transparency and repro-ducibility, preregistration may have the highest barrier to entry because it requires actions that are different from most researchers' current practices. It is common practice to 'plan' one's analysis strategy while conducting the analy-sis itself. This is fraught with risks of reasoning and reporting biases because analysis decisions can be influenced by the observation of the data.[24] In order to stimulate researchers' interest in trying out preregistration, COS launched the Preregistration Challenge[25, 26] a contest in which 1,000 researchers will earn US$1,000 each for publishing their preregistered studies. The Challenge is designed as an education campaign to increase awareness and knowledge about preregistration, and initiate the behavior to possibly instill preregistration as a habit. In sum, COS develops and evaluates a variety of initiatives to strengthen the incentives for open and reproducible practices among active researchers.

Infrastructure

Supporting all of COS's initiatives is the infrastructure to make transparency easy and practical. The Open Science Framework[27] is an open source frame-work to connect the services that researchers use, and provide an easy means of storing, archiving, preserving, and sharing one's research data, materials, and workflow. Researchers use the OSF as a virtual workspace to manage their pro-jects with their collaborators. By default, the projects and materials associated with them are private – available only to those individuals that the researcher designates. At any point, the researchers can choose to make parts or all of their project publicly accessible. The flexibility integrates the researchers public and private workflows and removes the practical barriers to openness, replacing them with only the question of whether one wishes to share. Simultaneously, the OSF offers tools to preregister studies, share pre-prints of articles or other research objects, and connect the storage services (e.g., Dropbox, GitHub, insti-tutional repositories) and other tools that researchers use (e.g., citation man-agers, analysis tools, data collection mechanisms, publication systems) into a single environment. Coupled with COS projects like SHARE[28] – an effort to create an open dataset of all research content – we are creating public goods infrastructure that supports the entire research lifecycle.

Cultural Change

COS aims to provide support for evidence-based changes that will align sci-entific practices with scientific values, and ultimately improve the efficiency of knowledge accumulation and its application to advancing the social good. COS has big goals, but cannot accomplish them itself. Collaboration and collective action are essential. The design of COS as a non-profit developing exclusively free, open-source tools means that COS has no competitors. COS can collaborate with and support others working toward similar ends without concern about competitive disadvantage.

Cultural change is a coordination problem. Many stakeholders need to contribute to shifting cultural norms in concert; initiatives pursued in isolation will falter. All of COS's initiatives depend on the broader research community supporting, embracing, and even driving changes to research practices and the culture of incentives more broadly. To the extent that COS has succeeded so far, it is largely a consequence of leaders and upstarts in the community supporting the efforts. For example, during the leadership of Eric Eich, *Psychological Science* initiated a variety of new initiatives including adopting badges for open practices. Also, via leadership of Alan Kraut, Sarah Brookhart, Bobbie Spellman, and many others, the *Association for Psychological Science* has supported COS efforts – such as adopting Registered Reports at Perspectives on Psychological Science – and other cultural changes. More generally, many researchers share our desire for change, and have been acting on their own, in collaboration with others, and in connection with COS.

The only competition that COS faces is the power of inertia and the status quo. Cultural change is hard, even when everyone agrees that change is needed. However, as 2016 begins, my simple assessment is not 'Change is possible' or 'Change is coming,' it is 'Change is happening.' Across the sciences, funders, journals, societies, universities, and researchers are taking on the task of identifying, testing, evaluating, and implementing changes to research practice to improve the quality and efficiency of research. Incentives are shifting toward embracing transparency and reproducibility as evidence of good practice.

But change is not complete. Researchers, particularly early-career ones, may find the uncertainties in the shifting culture unsettling. I find it exciting and liberating. The culture of incentives had previously been stacked against what most researchers believe is good practice and best for science. Now, the door is open for change and stakeholders across the scientific community are supporting that change.

We worry about kids losing interest in science as a cost to advancing knowledge and having an informed citizenry. To solve the problem, we look for ways to change their minds, and most efforts aren't working. Perhaps we instead need to focus on changing ourselves. I almost killed Haven's budding interest in science by describing how it actually works. As those realities shift, I can cultivate Haven's interest by showing how those scientific ideals that she is learning in 3rd grade are borne out in the daily practice of scientists around the world.

Acknowledgements

Preparation of this chapter was supported by the Laura and John Arnold Foundation, John Templeton Foundation, and the National Institute of Aging.

Notes

[1] Anderson Martinson & Vries, 2007.
[2] Iqbal, Wallach, Khoury, Schully & Ioannidis, 2016; Miguel et al., 2014; Wicherts, Borsboom, Kats & Molenaar, 2006.
[3] Button et al., 2013; Ioannidis, Munafo, R, Fusar-Poli, Nosek & David, 2014; Open Science Collaboration, 2015; Simmons, Nelson & Simonsohn, 2011.
[4] Nosek, Spies & Motyl, 2012.
[5] Button et al., 2013.
[6] Implicit Harvard, n.d.
[7] Dataverse, n.d.
[8] Open Science Framework, n.d.
[9] Open Science Collaboration, 2012.
[10] Open Science Collaboration, 2015.
[11] Errington et al., 2014.
[12] Ebersole et al., 2016; Klein et al., 2014.
[13] Silberzahn et al., 2016.
[14] Kidwell et al., 2016.
[15] Open Science Collaboration, 2016.
[16] Kidwell et al., 2016.
[17] Open Education Badges, n.d.
[18] Open Education Reports, n.d.
[19] Chambers et al., 2014.
[20] Nosek & Lakens, 2014.
[21] Nosek et al., 2015.
[22] TOP Guidelines, n.d.
[23] Wagenmakers et al., 2012.
[24] Gelman & Loken, 2014; Simmons, Nelson & Simonsohn, 2011.
[25] Preregistration Challenge, n,d,
[26] COS, n.d.
[27] Open Science Framework, n.d.
[28] SHARE, n.d.

References

Anderson, M. S., Martinson, B. C., De Vries, R. (2007). Normative dissonance in science: Results from a national survey of U.S. scientists. *Journal of Empirical Research on Human Research Ethics, 2,* 3–14. DOI: https://doi.org/10.1525/jer.2007.2.4.3

Button, K. S., Ioannidis, J. P. A., Mokrysz, C., Nosek, B. A., Flint, J., Robinson, E. S. J., & Munafo, M. R. (2013). Power failure: Why small sample size undermines the reliability of neuroscience. *Nature Reviews Neuroscience, 14,* 1–12. DOI: https://doi.org/10.1038/nrn3475

Chambers, C. D., Feredoes, E., Muthukumaraswamy, S. D., & Etchells, P. J. (2014). Instead of "playing the game" it is time to change the rules: Registered Reports at AIMS Neuroscience and beyond. *AIMS Neuroscience, 1*, 4–17. DOI: https://doi.org/10.3934/neuroscience2014.1.4

Dataverse. (n.d.). Available at https://dataverse.harvard.edu/dataverse/bnosek

Ebersole, C. R., Atherton, O. E., Belanger, A. L., Skulborstad, H. M., Allen, J. M., Banks, J. B., Baranski. B., Bernstein, M. J., Bonfiglio, D. B. V., Boucher, L., Brown, E. R., Budiman, N. I., Cairo, A., Capaldi, C., Chartier, C. R., Cicero, D. C., Coleman, J. A., Conway, J., Davis, W. E., Devos, T., Dopko, R. L., Grahe, J., German, K., Hicks, J. A., Hermann, A., Humphrey, B., Johnson, D. J., Joy-Gaba, J., Juzeler, H., Klein, R. A., et al. (2016). Many Labs 3: Evaluating participant pool quality across the academic semester via replication. *Journal of Experimental Social Psychology, 67,* 68–82.

Errington, T. M., Iorns, E., Gunn, W., Tan, F., Lomax, J., & Nosek, B. A. (2014). *An open investigation of the reproducibility of cancer biology research. eLife*, 3:e04333. DOI: https://doi.org/10.7554/eLife.04333

Gelman, A., & Loken, E. (2014). The statistical crisis in science. *American Scientist, 102,* 460. DOI: https://doi.org/10.1511/2014.111.460

Ioannidis, J. P. A., Munafo, M. R., Fusar-Poli, P., Nosek, B. A., & David, S. P. (2014). Publication and other reporting biases in cognitive sciences: Detection, prevalence, and prevention. *Trends in Cognitive Sciences*, 18, 235–241. DOI: https://doi.org/10.1016/j.tics.2014.02.010

Iqbal, S. A., Wallach, J.D., Khoury, M.J., Schully, S.D., Ioannidis, J. P. A. (2016). Reproducible Research Practices and Transparency across the Biomedical Literature. *PLoS Biology, 14(1)*, e1002333. DOI: https://doi.org/10.1371/journal.pbio.1002333

Kidwell, M. et al. (2016). Offering badges increases availability of research materials and data. Unpublished manuscript.

Klein, R. A., Ratliff, K. A., Vianello, M., Adams, R. B., Jr., Bahník, Š., Bernstein, M. J., Bocian, K., Brandt, M., Brooks, B., Brumbaugh, C., Cemalcilar, Z., Chandler, J. J., Cheong, W., Davis, W. E., Devos, Theirry, Eisner, M., Frankowska, N., Furrow, D., Galliani, E. M., Hasselman, F., Hicks, J. A., Hovermale, J. F., Hunt, S. J., Huntsinger, J. R., Ijzerman, H., John, M-S., Joy-Gaba, J., Kappes, H., Krueger, L. E., Kurtz, J. (2014). Investigating variation in replicability: A "many labs" replication project. *Social Psychology, 45,* 142–152. DOI: https://doi.org/10.1027/1864–9335/a000178

Miguel, E., Camerer, C., Casey, K., Cohen, J., Esterling, K. M., Gerber, A., Glennerster, R., Green, D. P., Humphreys, M., Imbens, G., Laitin, D., Madon, T., Nelson, L., Nosek, B. A., Petersen, M., Sedlmayr, R., Simmons, J. P., Simonsohn, U. Van der Laan, M. (2014). Promoting transparency in social science research. *Science*, 343, 30–31. DOI: https://doi.org/10.1126/science.1245317

Nosek, B. A., Alter, G., Banks, G. C., Borsboom, D., Bowman, S. D., Breckler, S. J., Bucj, S., Chambers, C. D., Chin, G., Christensen, G., Contestabile, M., Dafoe,

A., Eich, E., Fresse, J., Glennerster, R., Goroff, D., Green, D. P., Hesse, B., Humphreys, M., Ishiya, J., Karlan, D., Kraut, A., Lupia, A., Mabry, P., Madon, T., Malhotra, N., Mayo-Wilson, E., McNutt, M., Miguel, E., Levey Paluck, E., et al (2015). Promoting an open research culture. *Science, 348*, 1422–1425. DOI: https://doi.org/10.1126/science.aab2374

Nosek, B. A., & Lakens, D. (2014). Registered reports: A method to increase the credibility of published results. *Social Psychology, 45*, 137–141. DOI: https://doi.org/10.1027/1864–9335/a000192

Nosek, B. A., Spies, J. R., & Motyl, M. (2012). Scientific utopia: II. Restructuring incentives and practices to promote truth over publishability. *Perspectives on Psychological Science, 7*, 615–631. DOI: https://doi.org/10.1177/1745691612459058

Open Science Collaboration. (2015). Estimating the Reproducibility of Psychological Science. *Science, 349*(6251), aac4716. DOI: https://doi.org/10.1126/science.aac4716.

Open Science Collaboration. (2012). An open, large-scale, collaborative effort to estimate the reproducibility of psychological science. *Perspectives on Psychological Science, 7*, 657–660. DOI: https://doi.org/10.1177/1745691612462588

Open Science Framework. (n.d.). Available at http://osf.io/

Open Science Framewordk Badages. (n.d.). Available at https://osf.io/tvyxz/wiki/home

Open Science Framework Reports. (n.d.). Available at https://osf.io/8mpji/wiki/home/

Project Implicit. (n.d.). Available at http://implicit.harvard.edu/

Preregistration Challenge. (n.d.). Available at http://cos.io/prereg/

SHARE. (n.d.). Available at http://osf.io/share/

Silberzahn, R., Uhlmann, E. L., Martin, D. P., Anselmi, P., Aust, F., Awtrey, E. Bahnik, S., Bai, F., Bannard, C., Bonnier, E., Carlsson, R., Cheung, F., Christensen, G., Clay, R., Craig, M. A., Dalla Rosa, A., Dam, L., Evans, M. H., Flores Cervantes, I., Fong, N., Gamez-Djokic, M., Glenz, A., Gordon-McKeon, S., Heaton, T. J., Hedros Eriksson, K., Heene, M., Holfelich Mohr, A. J., Hogden, F., Hui, K., Johannesson, M., et al. (2016). *Many analysts, one dataset: Making transparent how variations in analytical choices affect results.* Retrieved from https://osf.io/j5v8f/

Simmons, J. P., Nelson, L. D., & Simonsohn, U. (2011). False positive psychology: Undisclosed flexibility in data collection and analysis allows presenting anything as significant. *Psychological Science, 22*, 1359–1366.

TOP Guidelines. (n.d.). Available at http://cos.io/top

Wagenmakers, E. J., Wetzels, R., Borsboom, D., van der Maas, H. L., & Kievit, R. A. (2012). An agenda for purely confirmatory research. *Perspectives on Psychological Science, 7*, 632–638.

Wicherts, J.M., Borsboom, D., Kats, J., & Molenaar, D. (2006). The poor availability of psychological research data for reanalysis. *American Psychologist, 61*, 726–728. DOI: http://dx.doi.org/10.1037/0003–066X.61.7.726

Open Course Development at the OERu

Wayne Mackintosh

OER Universitas (OERu) & OER Foundation, wayne@oerfoundation.org

Editors' Commentary

As the Massive Open Online Courses (MOOCs) produced by Ivy League institutions and driven by venture capitalists continue to pivot away from promises of serving the underserved, an international network of 30+ like-minded institutions known as the OERu have been quietly taking steps towards offering a free first year program of study for learners anywhere in the world. In this chapter, author Wayne Mackintosh describes the principles and processes of open course design and development that serve as the foundation of the network and its goal of providing free, open, flexible, student-centered, credit bearing, online education.

> *'I was excited to be offered something different and to be part of a pilot project,' reports Michelle Aragon, who in 2014 made history by becoming the first student to receive academic credit for completing a course through the OERu. Using free, open educational resources (OERs), and without leaving her home in Penticton in British Columbia, Canada, Michelle wrote a paper on child poverty in the Philippines and created a travel brochure on Bali's eco-tourism industry. Furthermore, as an OERu course, the only fees were to have her assignments and exams graded. 'I do feel a course like this requires a higher level of technological and research skills,' she adds. 'The expectation is for the student to access open resources online. That can be challenging but it's part of what makes taking this course a great experience.'*

How to cite this book chapter:
Mackintosh, W. 2017. Open Course Development at the OERu. In: Jhangiani, R S and Biswas-Diener, R. (eds.) *Open: The Philosophy and Practices that are Revolutionizing Education and Science.* Pp. 101–114. London: Ubiquity Press. DOI: https://doi.org/10.5334/bbc.h. License: CC-BY 4.0

Aragon, then a student in Thompson Rivers University–Open Learning's General Studies Diploma program, enrolled in AST1000: Regional Relations in Asia and the Pacific, a course created by OERu partner, the University of Southern Queensland (USQ) in Australia. USQ implemented a course design based on a pedagogy of discovery whereby learners identify OERs in pursuit of their own interests in achieving the learning outcomes for the course. 'This was an exciting opportunity for me as facilitator,' says Dr Marcus Harmes from USQ, 'and I believe for Michelle as learner to take an approach to study that allowed her to follow personal interests and areas of passion – the outcomes were excellent.'[1]

Introduction

Open course design and development at the OER *universitas* (OERu) flows from the culture of openness embedded into the OER Foundation (OERF). Consequently, any discussion about open course development must be situated and understood within the organisational context of the OER Foundation.

This chapter begins with a summary of the OER Foundation including its history, values, and its distinctly open operations. This background discussion will be followed by a succinct introduction of the OERu international innovation partnership. This discussion will provide the framework for reflecting on the open design and development practices at the OERu.

Organisational context of the OER Foundation

The Open Education Resource Foundation (OERF) is an independent not-for-profit organisation that provides leadership, international networking and support for educational institutions to achieve **their** strategic objectives using open education approaches.

Words are important

First, a note about nomenclature. We are **not** the 'Open Educational Resources' Foundation. For us, 'open education' is an umbrella concept encompassing multiple dimensions of openness including Open Educational Resources (OER), Open Educational Practices (OEP), Open licensing, open policy, free and open source software (FOSS), and open philanthropy. Resource (singular) is used as a noun to infer that openness is the primary means and enabler to achieve more sustainable education futures for all. Openness is the DNA of the OERF – we do not do closed as a matter of policy.

Conceived from failed innovation and organisational closedness

A strategy innovator by nature, I established the OER Foundation out of frustration with the slow rate of progress combined with the lessons learned from failed innovation attempts to transform educational institutions towards the mainstream adoption of open education approaches to leverage the affordances of a digital age in widening access to educational opportunity. The most important lesson learned was that the inertia of existing operations and business practices at many organisations is frequently too strong to achieve the critical mass required for substantive transformation towards open practices (see also Chapter 13 by Farhad Dastur).

To illustrate this point, I am a committed advocate and user of free software for education. Previously I was recruited to join an institution working in the field of open learning with full disclosure of my preferences for using FOSS. When it came to acquiring a new laptop on joining the institution, I suggested that I would use my existing machine which was customised with my personal flavour of the GNU-Linux operating system. I was informed that this would not be possible because the enterprise was required to 'maintain the image' which I subsequently found out was the Microsoft desktop image. It is a contradiction in terms to espouse open learning but to demand that individuals must sacrifice their freedoms in technology choice by requiring the use of proprietary technology.

I do not use non-free software as a matter of personal choice and my reticence to use non-free software was escalated to executive level. I received official notification that as an exception, I would be permitted to install a GNU-Linux operating system on an external hard drive with the enterprise issue of the Microsoft operating system installed on the laptop. I responded in writing that I would accept this requirement only if every other staff member in the organisation was required to have GNU-LINUX installed on their notebooks and provided with permission to boot Microsoft from an external hard drive. Fortunately, the organisation had the foresight to see the absurdity of discriminating against open systems and allowed me to run an open source operating system on the corporate notebook on condition that I did not generate support dependencies using a different system which was a fair and reasonable solution.

Cooperative independence as strategy for open transformation

Over two-decades of focused effort in attempting open transformation from within organisations, in the absence of satisfactory progress, it became clear that perhaps substantive transformation for openness could be better achieved through an independent organisation cooperating with existing educational institutions in the formal sector. I don't buy into the rhetoric of 'disruptive innovation' that universities are doomed to extinction. Universities are important organs of society. They are one of only a handful of organisations that have survived the industrial revolution and, if history repeats itself, they will survive deep into the knowledge revolution.

Early in 2009, we established the OERF as an independent charity. When searching for a suitable home for the Foundation, the best choice was to locate the new institution at Otago Polytechnic in New Zealand because they were the first tertiary institution in the world to adopt a Creative Commons Attribution intellectual property policy and the Council of the Polytechnic had the courage to register the OERF as an independent charitable entity.

The inkling that cooperative independence would be more effective in nurturing the development of sustainable ecosystems working towards the mainstream adoption of open education in the formal sector is paying dividends in the case of the OERF. As of 2016 the Foundation has no accumulated debt and its membership continues to grow at a steady pace. Currently the network comprises over thirty institutions from Africa, Europe, the Middle East, North America, Oceania, and South-East Asia.

Modelling the OERF structure on successful open source software foundation structures

The OERF is modelled on the organisational structures derived from successful open source software foundations like the Apache Software Foundation[2] and the Mozilla Foundation.[3]

The OERF, governed by an international Board of Directors, provides the legal framework for coordinating a number of flagship initiatives which operate as independent community projects. The OER Foundation is a social enterprise whereby money earned through our projects are invested back into the charitable education activities of the Foundation.

Meritocracy is a guiding principle of the OERF. Leadership roles in our community projects are 'earned' through sustained performance. Individuals who have gained respect from their community peers through engagement have a greater influence on decision-making. Transparent planning promotes trust in our open decision-making practices.

Flagship initiatives

The OERF administers a number of flagship initiatives in open education: OERu, WikiEducator, and hosts Creative Commons Aotearoa New Zealand.

The OERu is an international innovation partnership of universities, colleges, and polytechnics who are working together to provide more affordable and accessible higher education. WikiEducator is an international community of 60,000+ educators collaborating on the development of OER. CCANZ is the New Zealand national affiliate of Creative Commons and operates as a self-funded project.

Openness as principle not an optional feature

The OERF subscribes to the principles of open philanthropy and open governance for its operations and projects. Open philanthropy promotes radical transparency, sharing, and collaboration to effect social change in education. The OERF supports and encourages autonomy and open governance of its flagship projects. This enables the OERF to provide a clear distinction between legal and financial governance and the community-based operations of our flagship initiatives while providing the agility for individual projects to mature utilizing shared infrastructure. This networked model provides a low risk, low cost, but high impact innovation platform for institutions wanting to engage with open education.

Open planning

All planning activities of the Foundation's flagship initiatives are conducted openly and transparently. For example, the planning activities of the OERu project are hosted on WikiEducator with the freedom for any member of the public to help shape our futures. All partner meetings since inception of the OERu are streamed live on the internet encouraging wide international participation. The OERu project uses an open consensus model for decision-making and members from the open community also participate through the wiki and open email lists.

Open policy

As a general practice informed by the core values of the institution, the OERF staff do not participate in projects in their official capacity where the outputs are not licensed under a free cultural works approved open license. The free cultural works definition is derived from the essential freedoms associated with the free software movement.[4] So for example, restricting derivative works or commercial activity are a deemed to be material restrictions of freedom. Moreover, free cultural works approved artefacts must be stored in editable and open file formats. With reference to the suite of Creative Commons licenses discussed in Chapter 3, the Attribution and Attribution ShareAlike licenses and works dedicated to the public domain meet the requirements of free cultural works. In addition, we encourage that funding proposals are developed transparently and endorsements or participation from the OERF in philanthropic partnerships prefers that these documents are openly licensed. While some competing for contestable funding in open education are uncomfortable sharing proposals under open licenses, we at the OERF believe that if anyone 'steals' our ideas and can do what we propose quicker, cheaper or of better quality – then they deserve the funding. When outputs are released openly, as in the case of OER, we all benefit and the ecosystem grows. Requiring open licensing for joint funding proposals developed openly is also a great way to discern intent. We are sometimes approached by organisations to endorse or participate in peripheral capacities in joint proposals as mechanism to use the association with the OERF as an attempt to boost the likelihood of funding success.

Organisations who are unwilling to work openly care less about open than we do and are therefore not likely to be productive partners.

Open technology

As an open education project, the OERF uses FOSS exclusively for our enterprise infrastructure and we promote the use of open file formats. We use the same software as the popular Wikipedia website with a proven track record for security, reliability and scalability. Apart from the significant cost efficiencies gained from using FOSS, our choice is a values-based decision. In this way we can ensure that no educator in the world is restricted from participating in OER because they have to purchase software licenses or sacrifice their freedoms in software choices.

Designing for fiscal sustainability

I trained as an accountant in my first life (and do not readily admit this publically). However, this background has been a tremendous asset in establishing foundations for fiscal sustainability. I'm in the business of raising money so that the foundation does not make profit. That's an order of magnitude harder to do than running a successful commercial business. We decided to build the OERF from a very low cost base of less than US$200,000 per annum operating with only two full time staff members for the first 5 years. While this restricted our ability for rapid growth, we saved a ton of money which is now paying dividends because unlike a number of the commercial MOOC providers we do not need to figure out how to pay back millions of dollars worth of venture capital. The other lesson I have applied religiously is to avoid external funding to cover basic operations but to invest donor funding wisely in building strategic capability. In this way we have avoided generating too much dependency on third party donor funding.

The OERu international innovation partnership

'The OERu envisions a world where all learners have affordable access to higher education'

The OER universitas (OERu) is a consortium of over thirty post-secondary institutions and organisations (as of March 2016) collaborating on the assembly of university-level courses from OER and providing pathways for learners to achieve formal academic credit towards credible credentials. Coordinated by the OERF, the OERu is an international innovation partnership with member institutions from Africa, Asia, Europe, the Middle-East, Oceania, and North America.[5] Through the community service mission it is possible for organisations to invest time and effort to assemble courses based solely on OER. As accredited institutions, these universities, colleges and polytechnics can provide summative assessment services with pathways for learners to earn formal academic credit and pay reduced fees for assessment and credit when compared to full tuition. By combining the potential of OER with the community service mission, it is possible to create what Taylor (2007) has called a 'parallel universe' of post-secondary learning opportunities to complement and augment formal education provision, especially for those who lack the means to follow traditional learning paths. So for example, sharing course materials funded for mainstream delivery under an open license does not add additional cost for this institution if these are shared with the communities our public funded institutions are established to serve. On the contrary, this enables the organisation to serve a wider community without increasing cost. *Figure 1* below illustrates the OERu model which is designed to provide more affordable access to higher education leading to formal academic credit.

Originally conceived as the OER university (OERu) by the participants at the inaugural meeting of interested institutions in February 2012 we always used the lower case 'u.' The lower case 'u' refers to a community of scholars sharing information freely as intended by the original Latin meaning of *universitas magistrorum et scholarium* from which the word university was derived. In our

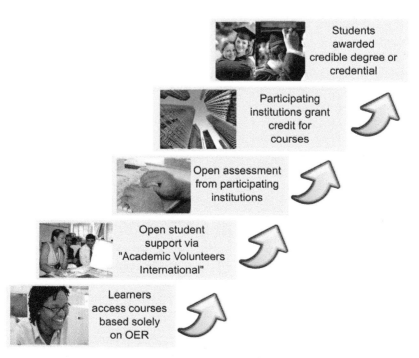

Fig. 1: The OERu Model.

case, the word 'university' did not refer to the title of formal teaching institution. The name was changed to OER universitas (OERu) when a group of formal universities objected to the use of the term 'university' which is a restricted concept in the Higher Education Acts of a number of countries.[6] This was a signal that our philanthropic collaboration was coming of age given the interest to assert 'ownership' of the concept 'university.' The OERF Board of Directors approved the adoption of the name 'OER universitas' which in hindsight better reflects the developing nature of the OERu network with increased membership from non-teaching institutions and a growing number of universities, community colleges, and polytechnics joining the network.

Open design and development at the OERu

Open design refers to the creation and development of potentially meaningful learning experiences through open and transparent collaboration among course developers and peers using open educational resources, open educational practices, and open technologies.

OERu design and development begins with a simple premise that it is more productive and sustainable to reuse and remix existing resources than to create new ones from scratch. It requires an agile disposition to assemble learning pathways which utilise existing OER and open access resources to support the learner's journey in attaining the learning outcomes. The open design process is highly iterative. Unlike production-line models found at many open distance learning institutions which develop a 'master design plan' which provides detailed direction of the development, the OERu design process accepts that we are more open to iterative change as the development process progresses. It draws on the experiences of open source software development. Eric Raymond compared the differences between open and closed models of software development in his seminal text, *The Cathedral and the Bazaar* (1999). The *cathedral* represents the detailed planning and closed development of proprietary software, where users only get to see the functionality and features between major releases and the code developed between releases is restricted to an exclusive group of developers. The *bazaar* references an approach where all code is developed on the internet in view of the public. Raymond proposed that 'given enough eyeballs, all bugs are shallow' which he terms Linus's Law named in honor of Linus Torvalds who led the development of the kernel of the GNU/Linux open source operating system.

The OERu provides an example of a design and development model which is distinctively open. The entire process from initial course nominations, to preparing design blueprints and developing the course resources is conducted openly on the internet for all to see and participate in. This open approach facilitates more iterative design and development because the design documentation becomes a living document and the open education community can assist with peer review and refinements. So for example, the design blueprint for the OERu's Digital Skills for Collaborative Development (DS4OER) course is shared openly with learners wanting to learn how to develop open design frameworks. In this instance, the two lead developers of the course were experienced wiki collaborators so they were able to work with a very basic design expression which evolved as the course development matured, rather than attempting to produce a 'textbook' blueprint to meet corporate design requirements. Learners participating in the course can view the edit history to see how the design evolved over time in conjunction with the wiki discussions associated with each page. The design concepts can easily be copied and are being reused for a wide range of new OERu course developments.

Describing open design

The concept of *open design* extends the principles of openness beyond OER materials themselves to include open planning, open design and open development of courses. Open design refers to the dynamic processes for open collaborative design and development of open courses. It draws on the open source software development model to facilitate rapid prototyping and continuous feedback and improvement loops.

In contrast to course development by a sole individual or dedicated production team, the open design approach is characterised by:

1. *Participants and teams constituting themselves in self-selected roles using collaborative processes.* Anyone is free to volunteer and contribute to the process.
2. *A highly iterative design and development process*, where people with different skill sets including learning designers, subject matter experts, language editors, and technologists work simultaneously rather than using a production line model with discrete division of labour.
3. *A public record of all planning and communications.* For example, creating a node page for planning the development in a wiki and using the corresponding discussion pages or posts to public email lists with public access to the archives.
4. *Open collaborative authoring technologies* which maintain a detailed edit history.
5. *Group decision-making informed by rough consensus and running code*, a concept coined by David Clark, a computer scientist. In open design, this means that the active collaborators tap into the 'sense of the group' at a given time to prioritise practical implementation knowing that the open model facilitates continuous improvement. In a rough consensus model, a majority agreement (i.e., 51%) of all listed participants is not required. In open design, it is better to have a working draft than an elaborate master plan agreed by the majority.

Dimensions of open pedagogy at the OERu

An open course at the OERu requires that anybody should be able to access the course materials without the need for password access. Individuals must be free to reuse, revise, remix, redistribute, and even sell our open content. The OERF supports the United Nations Declaration of Human Rights including the right to earn a living from our open resources. Our commitment to FOSS and open file formats ensures that educators and learners will be able to retain digital copies of their work.

Unfettered access to our course materials recognises the potential learning value of being able to fail anonymously. This feature could be of value to indigenous learners, first in family university learners, and learners who may perform better without the time constraints associated with completion of traditional courses. We do not know the extent that OERu courses serve these categories of learners because in the absence of mandatory registration we do not track learner progress. However, our server statistics confirm that our open courses attract visits from a large number of individuals who prefer not to register for a course. (At the OERu, course registration is optional for learners who would like to receive instructions via email.)

OER enables designers to implement a 'pedagogy of discovery' whereby learners can be guided to source their own open resources in pursuit of their own interests in achieving the course outcomes. The growing inventory of OERs and open access materials available on the web provides the opportunity to develop courses using a 'free range' learning strategy where learners can customise the content to suit their own needs and interests within the context of a university-level course.[7]

The OERu implemented this 'free range' learning model with a prototype course: USQ's Regional Relations in Asia and the Pacific, the course taken by Michelle Aragon, described at the beginning of this chapter. The Asia Pacific region comprises over forty different countries, and it would not have been possible to prescribe a closed text covering this wide range of countries. Whereas Michelle's work was assessed by the University of Southern Queensland in Australia, she successfully applied her credit towards her credential at Thompson Rivers University in Canada becoming the OERu's first graduate. Reflecting on the power of the model, Michelle notes: 'It was also quite freeing not to be tied to a textbook and able to follow what I wanted to learn about and what I wanted to write about.'[8] The OERF promotes the 'domain of one's own' philosophy where individuals have the freedom to manage and control their own technology and content. OERu course design encourages learners to maintain their own course blogs rather than e-Portfolio systems administered by the OERF or trapping contributions within the learning management system. In this way, learners have control over their own learning artefacts and will retain access to the outputs of their learning long after the course is completed. In another example, the digital skills for collaborative OER development (DS4OER) course teaches learners how to administer their own open source blog sites using free cloud-based services. In this way the OERu can empower educators to publish online course websites which may help cash strapped institutions in the developing world who do not have the resources to host their own technology infrastructure.

Lessons learned from implementing the OERu

In summary, the success of the OERu collaboration to date has been supported by the following guiding principles:

1. *Responding to a compelling vision which is well aligned to the core values of the contributing institutions.* The vision of providing free learning opportunities for all students worldwide with pathways to achieving affordable degrees, especially for learners who are excluded from the privilege of a tertiary education is a compelling and worthy vision. This is well aligned with the community service missions of the contributing partner institutions.
2. *Open sourcing everything.* The OERu is distinctively open using open educational resources, open educational practices, open licensing, open source software, and open planning models. Apart from significant cost savings in providing central technology infrastructure, open and transparent planning builds trust for existing and prospective partner institutions. All partners can monitor developments in real time and participate in all aspects of the implementation of the OERu without excluding valuable volunteer contributions from individuals in the open community.
3. *Ensuring the decision-making autonomy of partner institutions.* A key principle of engagement in the OERu model is the institutional autonomy of partner institutions regarding all decisions relating to the assessment and accreditation of learning. Partner institutions will not jeopardise their institutional stature, brand or credentialing authority yet working collectively the network is able to achieve more than working alone.
4. *Generating a viable value proposition for capacity development in open approaches.* Without tangible benefits for contributing partners, there is no motivation for institutions to contribute. The OERu enables institutions to participate in an international network while responding to their community service mission. The OERu model enables partner institutions to build capability in open and collaborative design models in online learning while generating opportunities for reducing cost. For example, partner institutions could diversify curriculum offerings for traditionally low enrolment courses which would be too expensive to produce alone, but could easily integrate an OERu course into the curriculum for full-fee students without incurring any capital course development costs. So

for example, Otago Polytechnic approved a course 'Change with digital technologies in Education,' a Masters level course originally developed by the University of Canterbury, for inclusion in the Graduate Diploma in Tertiary Education. The courses serve two different markets and is a good example of how a course funded by taxpayer dollars can help serve a wider range of institutions and students.

5. *Avoiding the temptation to innovate on too many fronts simultaneously beyond the capacity of the economy and society to accept the new developments.* While the allure of innovating through technology is appealing, the higher education sector and the economy are traditionally conservative when it comes to the token value of a university degree. The OERu has restricted its primary innovation to using courses based on OER for formal academic credit, and has intentionally left the innovation, for instance, of new forms of credentials like open badges to other players the ecosystem who are arguably better positioned to achieve success.

6. *Minimising risk while maximising impact.* The OERu network is a low risk opportunity for partner institutions because institutional exposure is limited to the assembly of only two courses from existing OER. However, the collective network returns are significantly greater than the initial investment of individual partners because the open model facilitates reuse and remix. Our open model allows the freedom for partners to contribute more than the minimum. So for example, Otago Polytechnic has 'donated' the Graduate Diploma in Tertiary Education which is a full program exceeding the suggested two course contribution. Recently, five partner institutions have agreed to share the costs of an open source software developer. In short, these partners gain the benefits of a full-time staff member for a portion of the cost because all the code developed through this positions is shared as open source software. On the other hand, 'silent partners' who take longer with their own course contributions still contribute to the greater good because their membership fees assist in maintaining the central infrastructure,

7. *Guaranteeing recoupment of future operational costs of contributing partners.* The recurrent costs of providing assessment services in the OERu model are recouped on a fee for service basis thus minimising risk for contributing partners and generating opportunities for new revenue streams.

8. *Incremental design combined with rigorous strategic planning.* It is not possible to develop a detailed master plan for the medium term in a highly volatile and fast moving technology environment in higher education. Moreover, the complexities associated with the dynamics of an international network comprising institutions from six major regions of the world cannot reasonably be anticipated within a rigid master plan. The OERu focuses on incremental projects which are small enough to fail but sufficiently strategic to facilitate organisational learning for the network. In this way the OERu remains agile and responsive to changing needs.

These principles are not mutually exclusive and interact with each other as a dynamic ecosystem. We believe that the OERu model is sufficiently agile and flexible to enable individual partners to pursue their own priorities without compromising the collective goal of widening access to more affordable education. We continue to learn on this open journey.

I am personally very excited by the OERu network's decision to progress a 'free' first year of study leading to an exit award as 'Minimum viable product.' Our tireless work in building trust through open and transparent planning has paid off as demonstrated by the unanimous decision of the OERu partners to approve the credit accumulation and credit transfer guidelines. It is indeed possible to nurture meaningful cooperation as we return to the core values of the academy which is to share knowledge freely.

Previous publication

Selected paragraphs in this text are proudly remixed from the author's own openly licensed contributions to:

- The Digital skills for collaborative OER Development open course, Available online at: http://course.oeru.org/ds4oer/
- Open Education Resource Foundation Ltd. Annual Report 2015. In press.

Notes

[1] OERu, 2014.
[2] http://www.apache.org/foundation/how-it-works.html
[3] https://www.mozilla.org/en-US/foundation/
[4] Freedomdefined, 2015.
[5] OERu: Undated.
[6] OERF, 2013.
[7] Taylor, 2012.
[8] OERu, 2014.

References

Freedomedefined. (2015). *Free cultural works definition*. Retrieved from http://freedomdefined.org/Definition

OER Foundation (2013). *The OER university becomes OER universitas*. Retrieved from: http://oeru.org/news/the-oer-university-becomes-oer-universitas/

OERu (2014). *Meet Michelle Aragon – OERu's first graduate*. Retrieved from: http://oeru.org/news/meet-michelle-aragon-oerus-first-graduate/

OERu (Undated). *OERu partners*. Retrieved from: http://oeru.org/oeru-partners/

Raymond, E.S. (1999). *The Cathedral and the Bazaar: Musings on Linux and Open Source by an Accidental Revolutionary*. Sebastopol, CA: O'Reilly Media. (p. 30) Retrieved from https://books.google.com/books?id=F6qgF tLwpJgC&pg=PA30&hl=en#v=onepage&q&f=false

Taylor, J.C. (2007). Open courseware futures: creating a parallel universe. e-Journal of *Instructional Science and Technology (e-JIST)*, *10*(1), 4–9.

Taylor, J.C. (2012). *The Pedagogy of Discovery: Enabling Free Range Learning*. Invited address to the Seminar on Creating and Widening Access to Knowledge Universiti Sains Malaysia, Penang, 26 June.

From OER to Open Pedagogy: Harnessing the Power of Open

Robin DeRosa* and Scott Robinson

Plymouth State University

*rderosa@mail.plymouth.edu

Editors' Commentary

The unaffordability of education—whether in terms of tuition or textbooks—has undoubtedly made the 'free' element of Open a rallying cry. However, as the open education movement matures, the fulcrum of this discussion appears to be shifting from an emphasis on the adoption of open educational resources to an embrace of open educational practices. In this chapter, authors Robin DeRosa and Scott Robinson draw on a variety of examples to illustrate the empowering potential of open pedagogy, an approach in which students are not just consumers of content but active and visible participants in the construction of knowledge. The chapter concludes with a reflection on some of the challenges and lessons learned from engaging students in public scholarship.

Understanding the Value of Open

There is no question that Open Educational Resources can save students money, and there is no question that the cost of higher education can be prohibitive for many students, so lowering costs is a shared imperative for those of us who are committed to educational access. But lowering costs always has to be contextualized into larger goals about learning. For example, a struggling university might curtail library hours or lower the heat in classrooms in order to save

How to cite this book chapter:

DeRosa, R and Robinson S. 2017. From OER to Open Pedagogy: Harnessing the Power of Open. In: Jhangiani, R S and Biswas-Diener, R. (eds.) *Open: The Philosophy and Practices that are Revolutionizing Education and Science.* Pp. 115–124. London: Ubiquity Press. DOI: https://doi.org/10.5334/bbc.i. License: CC-BY 4.0

money; these cost-saving measures could make education more accessible and could lower tuition and fees for students, but they could also impede learning. In this sense, cost savings is a complex issue, and any effort to save students money needs to be weighed against a range of repercussions distinct from the financial bottom line. The fact that learning materials now exist in digital formats does not necessarily mean that these learning materials can compete with traditional printed textbooks or other analog tools in terms of helping students learn. We have probably all tried some new-tech way of doing some old-tech process and found that the old-tech way worked better. So what about OER make them a good choice for adoption in the classroom? What, aside from cost-savings, make them valuable to education?

OER are free, digital, easily shared learning materials. Though colleges will have to address hardware issues (it would be a mistake not to acknowledge that access to the online world has real costs, and that free materials can only be freely available when institutions assure provision for all students), the potential that OER have to lower skyrocketing textbook costs is promising. When you look at the majority of research and press about OER, they focus on the rising costs of textbooks and the phenomenal cost-saving potential of OER. Individual students could save thousands of dollars over the course of an academic degree; colleges and universities could save hundreds of thousands — even millions — for their student bodies. In addition, institutions stand to strengthen their own financial health as they improve retention and enrollment rates by committing to OER initiatives. So should faculty convert to OER because it's cheaper for students? Or because it can improve the financial health of our institutions? Should we adopt OER simply because the technology is available? Or is something larger at stake here?

First, we need a corrective to the definition in the previous paragraph, since it's not enough to say that 'OER are free, digital, easily shared learning materials.' To be 'open' as well as free, educational materials must carry an open license (usually a Creative Commons license), meaning that OER can be reused, remixed, revised, redistributed, and retained. In other words, OER are flexible, and they empower faculty and students to work together to customize learning materials to suit specific courses and objectives. It's the way that the learning materials respond to learners and teachers that makes OER exciting; what should really galvanize faculty with an interest in educational transformation are the possibilities for pedagogical change that OER make explicit.

Student-centered pedagogy is clearly in fashion at the moment. But what does it mean to call an educational experience 'student-centered'? In many cases, people seem to conflate student-centered pedagogy with a customer-service model aimed at student satisfaction. Often, we hear 'student-centered' trotted out in policy discussions aimed at eliminating bureaucratic obstacles for students (for example, making transferring credits between institutions easier), or in faculty conversations about teaching methods. In the latter, faculty talk about increasing class discussion and refocusing classroom dynamics away from traditional lectures and toward a more interactive model. But in many cases, these new 'student-centered' policies do little more than respond to market demand, and these 'student-centered' pedagogies do little more than acknowledge a baseline student voice as part of the course. How can OER offer a more robust vision for centering our students in their educational experience?

By replacing a static textbook — or other stable learning material — with one that is openly licensed, faculty have the opportunity to create a new relationship between learners and the information they access in the course. Instead of thinking of knowledge as something students need to download into their brains, we start thinking of knowledge as something continuously created and revised. Whether students participate in the development and revision of OER or not, this redefined relationship between students and their course 'texts' is central to the philosophy of learning that the course espouses. If faculty involve their students in interacting with OER, this relationship becomes even more explicit, as students are expected to critique and contribute to the body of knowledge from which they are learning. In this sense, knowledge is less a product that has distinct beginning and end points and is instead a process in which students can engage, ideally beyond the bounds of the course.

If texts — content — are at the heart of a course, and content is now shaped into a process that depends on learner engagement in order to function fully, then OER propel us into truly student-centered territory. This territory might more aptly be described as 'learner-centered' or even 'learner-directed' if we follow through on the open pedagogy towards which OER gesture. (For the purposes of this inquiry, this article defines 'open pedagogy' in a way that remixes and revises the complex definition of 'critical digital pedagogy' set forth by Jesse Stommel.)

OER make possible the shift from a primarily student-content interaction to an arrangement where the content is integral to the student-student and student-instructor interactions as well. What we once thought of as pedagogical accompaniments to content (class discussion, students assignments, etc.) are now inextricable from the content itself, which has been set in motion as a process by the community that interacts with it. Moreover, students asked to interact with OER become part of a *wider public* of developers, much like an open-source community. We can capitalize on this relationship between enrolled students and a broader public by drawing in wider communities of learners and expertise to help our students find relevance in their work, situate their ideas into key contexts, and contribute to the public good. We can ask our students — and ourselves as faculty — not just to deliver excellence within a prescribed set of parameters, but to help develop those parameters by asking questions about what problems need to be solved, what ideas need to be explored, what new paths should be carved based on the diverse perspectives at the table. Open pedagogy uses OER as a jumping-off point for remaking our courses so that they become not just repositories for content, but platforms for learning, collaboration, and engagement with the world outside the classroom.

To help explore how this theory can be put into practice, we will offer a few specific examples of how open pedagogy – in concert with OER or even distinct from it – can empower learners in a course.

Wikipedia Assignments: an example of open scholarship

According to Alexa Internet, Wikipedia is the sixth most-visited website in the United States. When the general public searches the internet for information, Wikipedia articles are generally at the top of the search results. The accuracy of the information on Wikipedia is dependent on the constant contributions, revisions and confirmations from diligent contributors. An article can receive 'Good Article' status by the Wikipedia community if it meets six criteria (well written, verifiable, broad in coverage, neutral, stable, and if possible, illustrated with images). At the time of this writing, of the nearly more than 5 million articles in English Wikipedia, only about 35,000 (>1%) have achieved a Good Article status or better. Much work is yet to be done. The Wiki Education Foundation provides resources to help educators involve students in advancing access to knowledge while building their digital literacy skills. Their model involves an 'assignment' that replaces a more traditional research paper where the task is to write or edit and improve a Wikipedia article; perhaps attaining Good or even Featured Article status. Instead of writing solely for their instructor and a grade, they are writing for the public. This is vastly different than the disposable assignment where students' work (and feedback from the instructor) often end up in the class recycling bin at the end of the semester. Rather than only being 'graded' by an instructor, they may have to respond to other Wikipedia readers and reviewers. Have they written clearly enough for a general audience? Does their writing have a logical flow? Have they supported their statements properly and soundly? These are some of the questions and skills that Jon Beasley-Murray and his students at the University of British Columbia confronted when they worked on this kind of a project for a class called 'Murder, Madness, and Mayhem: Latin American Literature in Translation.' Their goals (all of which were attained) included improving Wikipedia's coverage of selected articles on Latin American literature, submitting these articles to Wikipedia review processes, and increasing the number of featured articles in this area. In Beasley-Murray's words:

'I decided to include wikipedia as a central part of a course I was teaching in the belief that it was only by actively contributing to the encyclopedia that they would learn about its weaknesses, and also its strengths. And also with the idea that they would thereby, and perhaps rather incidentally, improve articles in a field (Latin American literature) in which in my experience wikipedia has been especially weak.'

'I liked the idea that students would be engaging in real world project, with tangible and public, if not necessarily permanent, effects. In the end, an essay or an exam is an instance of busywork: usually written in haste; for one particular reader, the professor; and thereafter discarded.'

'I'd like to think that it is teaching the students research skills and writing skills in what is very much a real world environment. They were set a medium- to long-term goal at the beginning of the semester, and were required to work collaboratively both within their own groups and with strangers in the public domain to plan how to achieve and deliver that goal. And their final product is to be a professional piece of work that will be viewed by many thousands of people, a resource that is in most cases the first port of call for future researchers, whether students like themselves or the any of the many millions from all over the world who visit wikipedia. Most of these articles are, after all, the top hit (or very close to it) in any internet search of the topic. By comparison, the usual essays and exams that we assign our students really are rather pointless busywork.'

To date, more than 22,000 students enrolled in >1,000 courses have participated in the Wikipedia assignments, collectively working on more than 37,000 articles. Professional bodies like the Association for Psychological Science and the American Sociological Association have called on their membership to participate in their own Wikipedia initiatives. These calls have been heeded, with faculty like Paula Marentette, Erik Olin Wright, and Martha Groom among the 97% of instructors who report that they would teach with Wikipedia again.

Noba Project Student Video Awards: an example using video

The use of video can be a highly effective way to communicate information. In 2011, Salman Khan of the Khan Academy urged educators to use video to 'reinvent education' by flipping instructor content online. An open pedagogy perspective invites *students* to be content creators. Willmott et al. (2012) found such learner-centered activities can inspire and motivate students when they are engaged with course content. The Noba Project (see Chapter 16) offers an annual competition in which students submit creative short videos which address one of Noba's suggested psychology topics or issues. The videos should help viewers understand and remember the concepts around the topic and must be three minutes or less. The US$10,000 in prize money is distributed among the top videos each year. The 'products' that students create are impressive. The 2015 Noba Student Video Award projects are free and 'open' for review and reuse under a Creative Commons license. Not only does an experience like this engage and empower the students who create the content, but the content then becomes part of the learning process for future psychology students.

The Class-Created Textbook: An example from Robin's literature course

In my English course on early American literature, I realized that students were paying close to US$100 for a textbook filled with literature that was virtually all in the public domain. Of course, they were also paying for a host of helpful edits to that literature: spelling updates, excerpting decisions, explanatory annotations. One summer before I was going to teach the course again, I posted a call to alums of the course and asked if anyone wanted to help me track down public domain versions of the texts. About ten students were interested in working on the project, since the idea that they could build a replacement no-cost text for future generations of students in their program was appealing to them. By the end of the summer, we had a solid skeletal framework of texts assembled into an eBook that we built using PressBooks, a free user-friendly WordPress platform that makes it easy to create books and publish them online. When the course began, new students in the class took on editing duties: updating spelling; excerpting selected longer texts; and adding front matter for each chapter which included discussion questions and some interactive video. These were students with virtually no familiarity with early American literature, but they were able to do this work better than I could have, since they were essentially producing a collection targeted at their exact demographic. In addition, they felt remarkably more attached to their course textbook given the fact that they, in essence, were its authors. We have only taken one course of students through the book so far, and much of it is rough, but we share it openly and look forward to improving it every time a new group of learners engages with it and offers updates, corrections, and additions.

Part of what we realized was fun about the open textbook was the way that students in the course could do the work of real scholars, and the way that they could imagine a connection to future learners who would engage with their work. To capitalize on the sense of connection to a scholarly community (both senior and junior to them), I introduced a tool called Hypothes.is into the mix. Hypothes.is is a web annotation tool that allows readers of any website to annotate the text on that site, and these annotations can be public, allowing for other readers to reply and engage with the annotations. Our open textbook allowed us to share our text but also open conversations around that text, and represent our reading as something that was, in very visible ways, an act of participation in a scholarly community. Students loved it, and the whole process, from creating the OER to revising it with students to annotating it publicly, was catalyzed by the idea of the open textbook.

The Crowd-Sourced Syllabus: An example from Robin's writing course

Creating or adopting an open textbook can allow for exciting interactions between students, content, and broader academic and non-academic publics. But once we commit to student-directed learning, we can use open pedagogy to rethink our courses beyond the textbook, from the ground up, in ways that can ultimately make a student's experience in our class more meaningful. In my Composition course this semester, I started by asking students what they felt like they needed to learn. Together, we crafted our course objectives, just the way my American lit students had crafted our textbook. Some of the objectives they developed were predictably connected to writing, but others surprised me: for example, they wanted to add objectives about time management and about engaging with current events. From there, students crafted their own writing assignments and tied them to the course objectives. After that, we worked out a grading process that would allow them to self-grade using rubrics that we agreed on. Finally, all work would be done on their own public websites (their 'ePorts'), which they would build from scratch, control themselves, and take with them after the class ended. In effect, the course is centrally student-directed. It's also unfolding in public, in a kind of triangle between my website which contains our syllabus and readings, their ePorts where they do their work, and Twitter, where we chat together and share ideas with others who are interested. Obviously, this course is a fairly radical enacting of the principles of 'open pedagogy,' and it is probably too extreme a model for many faculty in many courses, but it does show us that when begin to empower learners to engage with the course content in truly dialogic ways, we can envision new possibilities for every level of course design.

Working in Public: Lessons and challenges on the ground

While some students will surely long for the days where they could just open a textbook, memorize a list of facts, and recite that list on an assessment, most students seem truly thrilled to finally be participating in – rather than just absorbing – their educations. Of course, as teachers, we have to weigh this kind of student empowerment against the very real challenges of working in public in these ways. First of all, privacy concerns are not only legitimate, they are also sometimes a matter of life and death. One of our students had transferred to our university as part of an elaborate plan to go into hiding from her abusive boyfriend; there could have been deadly consequences had she used her real identity online. There are good ways that students can protect themselves while working in public, but often times the risks are different for different students, and instructors need to educate themselves about safety concerns, about big data and how it is used, and about how the tools they recommend protect or compromise student privacy.

In addition to safety and privacy, another challenge relates to the idea that students will be putting work into the public commons that might reflect poorly on them because it is not polished or sophisticated. This was very clear in Robin's Composition course, where students sometime struggled with basic literacy issues that made their online writing very rough. To us, this is less about a problem with the students or their work, and more about the way that the web has been underutilized as a workshop space. The reason that we changed the term for student websites from 'ePortfolio' to 'ePort' is that we wanted to de-emphasize the idea that only perfectly polished work belongs on the web, and instead suggest that the internet is far more powerful and helpful to us if we use it as *portal* through which we can communicate with others. We try to model this ourselves by sharing our work online before it is polished and complete, and we try to emphasize all writing and creation as an iterative, ongoing process that the web can facilitate. If they post work that is not perfect (and of course, they only post work that is not perfect), we all offer suggestions and comments – publicly – and they keep working to improve the piece. It is our belief that the 'future employers' that we are often fearful will penalize our students for their imperfect public work will overlook the imperfection in favor of the evidence of a student's ability to engage in the digital world for the benefit of improving their projects. But this is something we all have to do together: to change the web from a stale collection of rapidly-outdating artifacts of perfection to a living, growing collaborative space where new ideas are always developing.

The Power of Open: A concluding thought

If we think of OER as just free digital stuff, as products, we can surely lower costs for students; we might even help them pass more courses because they will have free, portable, and permanent access to their learning materials. But we largely miss out on the opportunity to *empower* our students, to help them see content as something they can curate and create, and to help them see themselves as contributing members to the public marketplace of ideas. Essentially, this is a move from thinking about OER as open textbooks and thinking about them as *opening* textbooks...and all sorts of other educational materials and processes. When we think about OER as something we *do* rather than something we find/adopt/acquire, we begin to tap their full potential for learning.

Previous publication

A previous version of this chapter appeared in Educause Review on November 9, 2015 (CC-BY 4.0).

References

Amado, M., Ashton, K., Ashton, S., Bostwick, Nera, V., Nisse, T., Randall, D. (2015). *Project management for instructional designers*. Retrieved from http://pm4id.org/

American Sociological Association. (2016). *American sociological association Wikipedia initiative.* Retrieved from http://www.asanet.org/about/wiki_Initiative.cfm

Association for Psychological Science. (2016). *Wikipedia initiative.* Retrieved from http://www.psychologicalscience.org/index.php/members/aps-wikipedia-initiative

Creative Commons. (2016). *About the licenses.* Retrieved from: https://creativecommons.org/licenses/

Dean, J. (2015). *Undergrad Shannon Griffiths on using hypothesis in the classroom.* [Web log post]. Retrieved from https://hypothes.is/blog/undergrad-shannon-griffiths-on-using-hypothesis-in-the-classroom/

DeRosa, R. (2013). Open pedagogy [PowerPoint slides]. Retrieved from http://www.slideshare.net/orbitdog1/open-pedagogy-for-elearning-pioneers

DeRosa, R. (2015). *The open anthology of earlier American literature.* Retrieved from http://openamlit.pressbooks.com/

DeRosa, R. (2016a). *Composition.* [Web log post]. Retrieved from http://robinderosa.net/composition-2/

DeRosa, R. (2016b). *#opencomp gets rolling: Students in an open-pedagogy-based composition course.* Retrieved from https://storify.com/actualham/opencomp-gets-rolling

DeRosa, R., & Robison, S. (2015). Pedagogy, technology, and the example of open educational resources. *Educause Review.* Retrieved from http://er.educause.edu/articles/2015/11/pedagogy-technology-and-the-example-of-open-educational-resources

Guess, A. (2007). *When Wikipedia Is the assignment.* Retrieved from https://www.insidehighered.com/news/2007/10/29/wikipedia

Hypothes.is. (2016). *Annotate with anyone, anywhere.* Retrieved from https://hypothes.is/

Jbmurray/Madness. (2016). *Was introducing Wikipedia to the classroom an act of madness leading only to mayhem if not murder?* Retrieved from https://en.wikipedia.org/wiki/User:Jbmurray/Madness

Khan, S. (2011, March). *Let's use video to reinvent education.* [Video file]. Retrieved from http://www.ted.com/talks/salman_khan_let_s_use_video_to_reinvent_education

Laurent, M. R., & Vickers, T. J. (2009). Seeking health information online: Does Wikipedia matter? *Journal of the American Medical Informatics Association, 16*(4), 471–479. Retrieved from http://jamia.oxfordjournals.org/content/16/4/471

Marentette, P. (2014). Achieving 'good article' status in Wikipedia. *Observer, 27*(3). Retrieved from http://www.psychologicalscience.org/index.php/publications/observer/2014/march-14/achieving-good-article-status-in-wikipedia.html

Noba Psychology. (2015). 2015 Noba student video award recipients. Retrieved from http://nobaproject.com/student-video-award/winners

Noba Psychology. (2016). *2016/17 Noba + Psi Chi Student Video Award.* Retrieved from http://nobaproject.com/student-video-award

Pawlyshyn, N., Braddlee, Casper, L., & Miller, H. (2013). *Adopting OER: A case study of cross-institutional collaboration and innovation.* Retrieved from http://er.educause.edu/articles/2013/11/adopting-oer-a-case-study-of-crossinstitutional-collaboration-and-innovation

Senack, E. (2015). *Open Textbooks: The Billion-Dollar Solution.* Retrieved from http://studentpirgs.org/sites/student/files/reports/The%20Billion%20Dollar%20Solution.pdf

Stommel, J. (2014). Critical digital pedagogy: A definition. *Hybrid Pedagogy. Retrieved from* http://www.digitalpedagogylab.com/hybridped/critical-digital-pedagogy-definition/

The Wiki Education Foundation. (2016a). *Inspiring learning. Enriching Wikipedia.* Retrieved from https://wikiedu.org/

The Wiki Education Foundation. (2016b). *How do you measure the difference that open knowledge makes?* Retrieved from https://wikiedu.org/changing/classrooms/

Whitman, W. (2015). Song of myself. In R. DeRosa (Ed.), *The open education anthology of earlier American literature.* Pressbooks. Retrieved from https://via.hypothes.is/http://openamlit.pressbooks.com/chapter/song-of-myself/

Wikipedia. (2016a). *Good article criteria.* Retrieved from https://en.wikipedia.org/wiki/Wikipedia:Good_article_criteria

Wikipedia. (2016b). *Wikipedia: WikiProject Murder Madness and Mayhem.* Retrieved from https://en.wikipedia.org/wiki/Wikipedia:WikiProject_Murder_Madness_and_Mayhem

Wikiwand. (2016). *Alexa Internet.* Retrieved from: https://www.wikiwand.com/en/Alexa_Internet

Wiley, D. (2013, October 21). What is open pedagogy? [Web log comment]. Retrieved from http://opencontent.org/blog/archives/2975

Willmot, P., Bramhall, M. & Radley, K. (2012). *Using digital video reporting to inspire and engage students.* Retrieved from http://www.raeng.org.uk/publications/other/using-digital-video-reporting

Wright, E. O. (2012). *Writing Wikipedia Articles as a Classroom Assignment.* Retrieved from https://www.ssc.wisc.edu/~wright/ASA/Writing%20Wikipedia%20Articles%20as%20a%20Classroom%20Assignment.pdf

Opening Up Higher Education with Screencasts

David B. Miller* and Addison Zhao†

* Department of Psychological Sciences, University of Connecticut,
david.b.miller@uconn.edu

† Department of Educational Psychology, University of Connecticut

Editors' Commentary

In this chapter, authors David Miller and Addison Zhao discuss opening the classroom with the use of screencasts. Screencasts are, in essence, video lectures. Their digital format carries all the benefits of that medium: the ability of students to learn from distance, the ability of students to learn asynchronously, and the ability to scale the classroom. In this sense, the authors are using the word 'open' in a somewhat different fashion than it is used elsewhere in this volume. Rather than speaking about open licensing or remixing of materials he is specifically referring to a process of expanding the boundaries of the classroom. Embedded within this, however, is a sentiment that is not far from the heart of open education: a lack of protectionist attitude. While the authors honestly and responsibly address the legal use of copyrighted information in screencasts they also directly acknowledge more open alternatives and their potential benefits.

What will a technologically-based classroom look like in 100 years? This is a question that was addressed by Jean Marc Cote in 1901 in a painting entitled, 'At School'.[1] This painting depicts a teacher grinding up books, with the aid of an assistant (possibly what might today be a Teaching Assistant), while six male students sit at tables wearing audio headphones that are connected by wires to

---

How to cite this book chapter:

Miller, D B and Zhao A. 2017. Opening Up Higher Education with Screencasts. In: Jhangiani, R S and Biswas-Diener, R. (eds.) *Open: The Philosophy and Practices that are Revolutionizing Education and Science*. Pp. 125–138. London: Ubiquity Press. DOI: https://doi.org/10.5334/bbc.j. License: CC-BY 4.0

the ceiling running off toward a wall with allegedly some means of transmitting audio information. One student, like some of today's students, is even staring out the window. This painting is rather prophetic in several ways in as much as the iPod was invented in 2001, and, with the ever-increasing costs of textbooks, professors have been examining alternative means of content delivery, such as electronic books, podcasts, and online interactive software.

In this chapter, we discuss issues concerning effective multimedia design and how to incorporate that into the creation and distribution of screencasts, which are movies often distributed as 'video podcasts' composed of narrated Microsoft PowerPoint or Apple Keynote presentations. My (DMB) own motivation in creating screencasts as a means of content delivery stems from my concern that information presented in a classroom is a momentary event. Students are unable to 'rewind' and easily revisit content that they were unable to grasp initially without asking the instructor to repeat the information. While such interruption and repetition does sometimes occur, it can interrupt the flow of content delivery and lead to disengagement by students who grasped the material initially. Screencasts enable students to control the pace of content delivery because movies can be made available throughout the semester for each student to watch at his/her own pace. Openly sharing course content throughout the semester can improve pedagogy by facilitating better note-taking for those students who 'rewind' the content delivery, as well as better understanding of material by enabling students to revisit those portions of the screencasts that may have been unclear upon initial viewing. Of course, sharing such materials in an open market beyond one's classroom can sometimes be challenging if materials are included that are copyright-protected, as we discuss later in this chapter. But, at least the materials are available at all times to students enrolled in the course.

The ability and ease of being able to share content either openly or in partially-open markets has coevolved with technological innovations that have been incorporated into classrooms throughout the century, but mostly in recent decades. While the future coevolution of such innovations is a matter of speculation, it might be informative to visit where it all began up to the present time.

Classroom Technology

Early forms of classroom technology were in analog format and not easily disseminated. Most likely, the earliest form was the chalkboard, which was considered to be a technological innovation when it began to appear in classrooms around the early 19th century.[2] Their popularity and enduring addition to classrooms might be attributed to the fact that they are inexpensive and relatively-low maintenance.

Electricity opened up new technological options, such as the overhead transparency projector, which evolved from non-electric 'magic lanterns' that were invented in the 15th century.[3] The modern version of the overhead transparency projector appeared in the mid-1940s, followed by the Kodak Carousel slide projector in 1961.[4]

Rapid advancements were made in classroom technology in the second half of the 20th century and in the early 21st century. Slide projectors were increasingly incorporated into teaching from the 1960s to the 1990s and were gradually replaced with computers and video projectors. By the 1990s, overhead transparency projects evolved into document cameras. Personal computers began to appear in the 1980s, which opened up new avenues of media creation and ease of distribution.

Every advancement in hardware has been closely followed by corresponding advancements in software. What has lagged behind, until recently, is research on the effective use of such technological innovations for the enhancement of teaching and learning. Digital technology has enabled us to share video, audio, and image creations almost seamlessly, which has empowered educators to incorporate varied multimedia formats into their classrooms. But, with great power comes great responsibility, especially when using presentation software.

Multimedia Design

The advent of presentation software, such as Microsoft PowerPoint, starting around 1990, gradually led to a decline in the use of slides and overhead transparencies in the classroom. Currently, high-tech classrooms rarely incorporate carousel-type slide projectors, and overhead transparency projectors have been replaced by document cameras. Apple Keynote, originally created for Apple's CEO Steve Jobs' own presentations, became publically available in 2003. Once Keynote had been upgraded in 2006, and especially in 2009, it became an ideal alternative to PowerPoint for reliably incorporating multimedia into presentations by Mac users.

The ease of use of PowerPoint and Keynote brought with it a means of misuse and outright abuse. Needless templates with distracting background designs and images decreased available screen space for important information. Easy-to-implement, distracting animations within and between slides became prevalent and, in most cases, unnecessary. But, most notable has been the tendency of presenters to convert what used to be lecture notes into endless lists of bulleted text. Some presenters have the tendency to simply read such bullet points line by line off the screen to their increasingly disengaged students. A 2009 study found that, 'the most important teaching factor contributing to student boredom is the use of PowerPoint slides.'[5] But, it is not so much the 'use' of presentation software that leads to student disengagement and, therefore, ineffective teaching and learning, but rather the 'misuse' of such software.

Fortunately, abusing presentation software in this manner can be remedied with some time and effort.[6] One approach is replacing bulleted text with

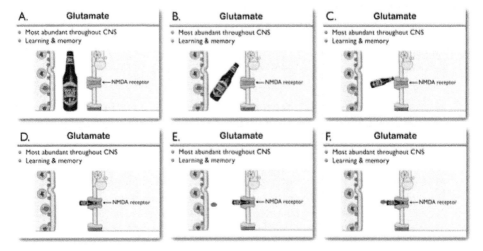

Fig. 1: Sequence of actual animation showing how alcohol blocks the action of the neurotransmitter, glutamate. 'CNS' is an abbreviation for central nervous system. (A.) Alcohol is portrayed as a beer bottle occupying the synaptic cleft between two neurons. (B.) Beer bottle shrinks and repositions as it moves toward NMDA receptor in neuron. (C.) Beer bottle continues to shrink and orients toward blocking NMDA receptor. (D) Beer bottle now totally blocks NMDA receptor. (E.) Glutamate, shown as a purple oval, is released from presynaptic neuron on the right and moves into the synaptic cleft toward so that it can lock into the NMDA receptor site. (F.) Glutamate cannot move into the NMDA receptor site because it is blocked by the beer bottle. The result is that learning and memory is disrupted because of disrupted communication in the hippocampus, which is a structure in each hemisphere of the central nervous system and is hugely involved in regulating learning and memory.

multimedia, typically with images and/or video clips. This replacement necessitates thinking about concepts and terms visually. For example, instead of a bullet point indicating that the amygdala is an area of the brain involved in emotion, show a sketch of the brain, highlighting the amygdala, surrounded by faces expressing various emotions. Or, instead of using bullet points to indicate that alcohol blocks the action of the neurotransmitter glutamate in the hippocampus, use shapes to draw a prototype receptor site, have an image of a bottle of beer, animate the beer image to reorient and move into the receptor site, preventing an image of glutamate from locking into this receptor site now blocked by the beer, and concluding to students how it disrupts the hippocampal regulation of learning and memory (see Figure 1).

Constructing slides in this manner helps the instructor think about visual means of portraying what otherwise would be expressed in bulleted text. Adding animation of objects wherever appropriate, and even sound effects, helps to convey the message.[7]

Screencasting

Following the popularity of the iPod and Apple's iTunes Store in the early 2000s, a new form of transmitting information began to gain popularity. Podcasting, as it was called when it emerged in 2004, was designated the 'word of the year' in the New Oxford American Dictionary in December 2005. Podcasting, defined as, 'A digital audio file made available on the Internet for downloading to a computer or portable media player, typically available as a series, new installments of which can be received by subscribers automatically,' changed the information delivery paradigm. Soon, some instructors began using audio podcasts as part of classroom assignments, but studies have shown that transmitting information via audio-only forms of communication is not very effective. For example, a 2010 study found that students who only receive instruction via podcasts perform more poorly than students who receive equivalent instruction via reading a text.[8] The researchers further suggest that podcasts might be useful as a course supplement or enrichment, but not as a means of primary content delivery. As prophetic as Jean Marc Cote (1901) was, the future does not lie in transforming the delivery of information from one source (reading) to another (audio).

In Fall 2005, I (DBM) began a podcast series called *iCube: General Psychology at The University of Connecticut*. (I shortened the name in 2014 to *iCube: UConn Psychology*.) The main component of that series was a weekly discussion with students from my large (n = 300–400 students) General Psychology course who chose to show up at a preset weekly time to ask questions and/or discuss topics related to what I had been covering in class. Unexpectedly, we soon had an international audience that resulted in interesting emails from listeners who wanted to learn more about psychology and, in some cases, coincidentally improve their English. *iCube* was, however, only a course enhancement. It was never intended as a primary means of content delivery. Even as a supplement, 10 years of anonymous course evaluation data showed that a majority of students (76%) who listened regularly noted that listening to the podcasts helped them learn.

In Fall 2006, I began another audio podcast series for the Honors students in my Animal Behavior course called, *Animal Behavior Podcasts*. That, too, drew listeners from around the world. Like *iCube*, it was a weekly discussion of animal behavior research. As a participant in *Animal Behavior Podcasts,* I (AZ) found the required discussions to be a meaningful avenue for expanding research ideas or questions that I had while learning the course content. In many ways, this discussion group resembled a community of learners who actively sought to take their understanding from remembering to analysis and future implications. For students interested in research, this discussion group allowed them to practice the scientific method of exploring the problem, finding the variables, and coming up with viable research approaches.

From Podcasts to Screencasts

Beginning in the Fall of 2009, I decided to explore ways of incorporating video into my Animal Behavior course. Not long before that, I became aware of screencasting as a means of recording everything on a computer screen (such as PowerPoint or Keynote presentations) while simultaneously recording narration over those screens, followed by using powerful tools for post-production editing. I was intrigued by the idea of converting my multimedia lectures into a video format that allowed students to access them day and night, seven days a week, while being able to pause, take detailed notes, replay any portions they were unclear about, and so on. Screencasting seemed like an ideal means of doing this, and I soon discovered software that was both easy to use and offered powerful editing capabilities–ScreenFlow.[9] An alternative for PCs is Camtasia (https://www.techsmith.com/camtasia.html), which also offers a version for Macintosh. Free alternatives exist, like RecordIt,[10] but they lack the powerful editing features found in ScreenFlow and Camtasia. Strictly for screen recording, Open Broadcaster Software on the PC allows users to record their computer's screen, computer audio, and additional audio from a microphone. For anyone who does not need the editing capabilities, this software works well to create videos, and an open-source editor could be OpenShot Video.[11] While I (AZ) have never used OpenShot Video, a quick search can let someone see that open source software for recording and editing exists. In addition to the software, I (DBM) used a Samson C03U USB microphone connected to my MacBook Pro laptop to do the recording. I also recorded the screencasts directly onto an external hard drive because the internal hard drive had limited space. The external hard drive also allowed simple backups and portable transfer when necessary.

It took over 400 hours to create 85 screencasts of varying length for this course, but the effort was well worthwhile. My lecture course now became a hybrid course, in which around 90% of the material was delivered via screencasts, and 10% in weekly live lectures consisting mostly of newer material that was published after the screencasts were recorded.

A notable facet of the Animal Behavior hybrid course was the consistency of the information I (AZ) received. Since the screencasts were produced similarly to the in-class lectures, I was familiar with the content delivery format even if I did not have logistical control (e.g., pausing, rewinding) over the material received in the lecture version. This consistency led me to assimilate information from the screencasts and the lectures without interruption. Often times, instructors look at hybrid courses as ways to deliver information one way online, and then use the class time for a separate instructional strategy (e.g., group work, presentations, etc.) but the Animal Behavior course was effective because the content delivery format was high quality, online, and in person. This method indicates that not all technology needs to be integrated in a special or unique way to be effective.

Distribution of Screencasts

Before implementing the Animal Behavior screencasts, I had to negotiate a means of secure distribution because of the large amount of copyrighted material. Such material included video clips from nature television programs, as well as videos and images from various online sources and publisher-supplied materials. It had been safe to project that material in a lecture because it had not been distributed to the students, but that would not be the case with screencasts. After a month of working with my university's Attorney General Office to get official permission to post these screencasts online, all parties agreed that this could be done by hosting the screencasts on my university's password-protected server after converting the videos to stream-only, rather than downloadable, and putting a notice on the website where the videos would (a) only be accessed by students enrolled in the course, (b) intended only for use in the course, and (c) could not be copied.

The conversion process was simple. I (DMB) used ScreenFlow to export the edited screencasts as QuickTime movies in .mp4 format. Then, our information technology staff converted those videos to streaming videos as reference movies that were linked to the actual videos on a streaming server. Students only had access to these streamed videos, and downloading them would lead to a reference link that would not be the actual videos. Thus, in this manner, the videos were protected from direct downloading.

Students' Responses to Screencasts

Student responses to questions that I added to our anonymous course evaluations revealed that most of the students favored this method of content delivery. Bear in mind that my screencasts had little in the way of bulleted text. Also, my narration was simply talking about what was onscreen rather than reading from a prepared script. I chose this format in order avoid the lack of spontaneity and boredom that might come from being read a script.

Figure 2 shows students' responses on a scale from 1 (Strongly Disagree) to 10 (Strongly Agree) to the statement, 'The online method of content delivery helped me learn.' Figure 3 shows their responses to the statement, 'If given a choice between two sections of this course, both taught by Dr. Miller, I would prefer this online version instead of a regular in-class version.' For each of these figures, the number of respondents ranged from 126 to 176 each year (see legend insert in Figure 4 for exact numbers).

Finally, Figure 4 shows the course grades for the hybrid course (2009 through 2014) compared to the last time I taught the in-class lecture course (2008). Exams were kept the same between 2008 and 2009, and similar each year thereafter. Course grades were based on two mid-term exams and a non-cumulative final exam.

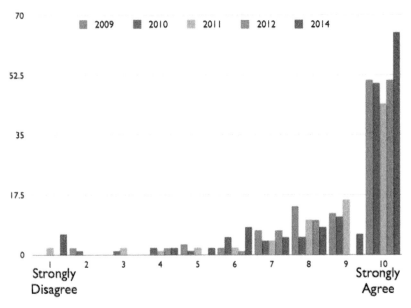

Fig. 2: End-of-semester, anonymous responses of Animal Behavior students to the statement, 'The online method of content delivery helped me learn.'

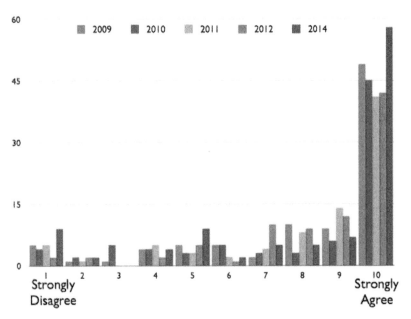

Fig. 3: End-of-semester, anonymous responses of Animal Behavior students to the statement, 'If given a choice between two sections of this course, both taught by Dr. Miller, I would prefer this online version instead of a regular in-class version.'

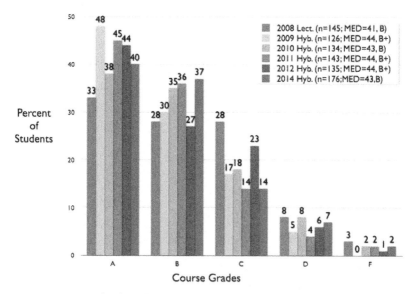

Fig. 4: Course grades based on two mid-terms and one final exam in Animal Behavior. The hybrid versions of the course from 2009 through 2014, which employed screencasts, resulted in better student performance than the 2008 in-class version. (The 2008 data are consistent with the many years prior to that in which the lecture version of the course had been taught.)

It is clear from students' responses as well as their grades that the screencasts were highly effective in student learning. Many anonymous written statements also supported the efficacy of the screencasts.

The success of the 2009 screencasts led me to do a major revision in the Fall of 2014. All of the original 85 screencasts were replaced with 84 newly-created videos for use in the course beginning in Fall 2015 and subsequent semesters.

Podcasting Revisited

I discontinued the *iCube: UConn Psychology* podcast series in the summer of 2015 after a successful 10-year run. In its place, I created a new video podcast series called *Psychological Science*.[12] Like *iCube*, the new series is intended primarily for the students in my General Psychology course. But, unlike *iCube*, this series consists only of screencasts that review, topic-by-topic, major points covered in the most recent lecture. These screencasts become available to students within a day after each lecture. To date, I have created 85 screencasts for this course.

A pilot program that I incorporated into *iCube* for several semesters indicated that students greatly valued having access to such screencasts, especially since I have not used a textbook in this course since around 1995. I never pre-pared lectures from textbooks, so they were never a meaningful pedagogical enhancement to my course. Moreover, the costs to students began to increase throughout the years, so I eliminated them altogether. Instead of a textbook, I created a student manual containing course policies, graphics that I pro-duced, a detailed course outline, and miniature versions of many of the slides (minus multimedia material) that I used throughout the course so that students could follow along with my lectures if they so choose. Another key to hav-ing the slides printed out ahead of time in a 'textbook' format is being able to align the course objectives and content with the material presented. By being on the same page as the professor every step of the way, students (like myself [AZ]) were able to know exactly what was taught, coming, and needed to be reviewed. This clarity helped to ensure that students would not be shuffling through information trying to find out what was necessary for the test, par-ticularly because the printed slides were incomplete and students were the ones to fill in those gaps. Because there is no textbook in my course, the screencasts enabled the students to review the material, revise their notes to include points they might have missed in class, and, for students who were absent, get a better idea of what they missed in class before borrowing notes from other students. By encouraging students to review material in a concise format on a regular basis, they are given additional avenues to assimilate the material instead of simply reviewing all their notes in one session before an exam. At UConn, class attendance is not mandatory, but I've always had very high attendance in my class throughout the semester. The availability of the screencasts did not affect attendance, and I noticed an increase in students earning A's in the course once these became available.

Other Uses for Screencasts

Following the creation of the initial set of Animal Behavior screencasts, I soon discovered other applications of screencasts that did not involve course-flip-ping, and that have been enormously useful. Because UConn is located in New England, we have many days when, due to weather (usually snow), all classes are cancelled. These cancellations wreak havoc on professors' syllabi as they find themselves getting far behind what they had intended to cover, resulting in continuous revisions of lesson plans, exams, and assignments as the semes-ter proceeds. Fortunately, I've never gotten behind in my lectures. Whenever the university closes due to weather, I record a screencast of that day's lecture, upload it to our server, and send the link to all of my students, alerting them that the next class will pick up where the screencast ends. Students have responded favorably to these screencasts because changes in class schedules can have a negative impact on students' plans. This is because they'd have to attend make-up lectures arranged by professors at times that they might not find convenient because of other commitments. At my university, the Registrar designates one or more Saturdays as make-up days, and students often have work and/or other responsibilities that would make it impossible to attend such sessions. Also, on the rare occasions when I have to be away from campus, such as giving a talk at a conference on a day that I would otherwise be lecturing, I record a screencast of the missed lecture prior to leaving town and, again, alert the students to view it before the next class.

I have also found it useful to create instructional screencasts for my graduate student course assistants and exam proctors. These screencasts enable me to effectively convey my expectations of their duties. Because of my large classes, I have worked out a somewhat convoluted means of assembling the exams in terms of putting seat numbers on each exam in a specific order such that students entering the room will not end up sitting next to one another. My screencast instructions for my course assistant enables me to show this exam-assembly process without ambiguity, and something he/she can refer to as needed throughout the semester.

Exam proctors often complain about professors not informing them about their duties. My screencast instructions for proctors allow me to explain in detail what I expect of them and how to go about proctoring my exams. I typically have up to 10 proctors for each exam, so being able to centralize the instructions via screencasts has been a huge time-saving mechanism.

I have also created, on request, and free of any Creative Commons or other license, screencasts on teaching tips for graduate-student 'instructors-of-record'.[13] These are graduate students who have been assigned to teach their own courses rather than simply serve as course assistants or teaching assistants. These screencasts have also been useful to newer faculty who have not had a great deal of teaching experience, and sometimes no experience, teaching large courses.

Others have suggested that professors might consider screencast assignments for students.[14] As Ledonne (2014) indicates, most students are comfortable with technology, and the creation of videos engages a number of learning styles (audio, visual) that would not occur in a traditional culminating assignment such as a term paper; moreover, students are more likely to find such a project more engaging than a term paper and, if successful, shareable in the form of an end-of-semester 'film festival' or, if appropriate, posting on YouTube for viewing by friends and relatives. Of course, it is important that students are trained in the proper use of such technology and that adequate resources are made available by the institution to enable students who might not have their own equipment to create such projects. In terms of training, Young (2011) pointed out that some colleges are considering including courses on video production as part of the curriculum. If an instructor is already creating screencasts for their course, a guide can be easily made showing the process for a course screencast, and the course content would serve as a constant model for producing effective videos.

Distribution of Screencasts in Open Education Commons

For effective consumption of screencasts, it is important that these videos be made available in standard, cross-platform formats. Depending on the software on one's own computer, the usual formats are .mov, .mp4, and .wmv files. In the case of iTunes and iTunes U, the videos are converted automatically for viewing on computers and mobile devices.

Additional to iTunes and iTunes U (for those universities that have contracts with Apple for hosting screencasts on iTunes U), other typical outlets include YouTube, Vimeo (both of which permit the explicit choice of a Creative Commons license), Google Drive, and Facebook. Screencasts can also be uploaded into one's Dropbox 'Public' folder, where the links can be shared with specific viewers. There are also sites that specialize in hosting videos for business, such as Wistia.[15]

Another option for sharing videos (as well as sounds and images) is on Wikimedia Commons,[16] which is licensed under Creative Commons. As such, any uploaded video must be one's own work, which might place certain limitations on distribution.[17]

Of course, the issue of incorporating copyrighted material within screencasts that are hosted on non-password-protected sites remains a problem for open education. Obtaining rights can be very time-consuming and, in some cases, monetarily costly. A workaround for images is to search for content on http://images.google.com, click on 'Search Tools' just above the image thumbnails, then click on 'Usage rights.' For educational markets, the best option to now choose from the dropdown menu is either 'Labeled for noncommericial reuse with modification' if you plan on altering the image, or 'Labeled for noncommercial reuse' if you do not plan on altering the image. The 'Usage rights' option is not available for video. There are, of course, other websites that have Creative Commons search engines such as Flickr,[18] the British Library,[19] and others.

A limited workaround for video in screencasts that will be uploaded to YouTube is to link to a video rather than embedding the actual video into Keynote. Once a screencast has been exported from ScreenFlow and uploaded to YouTube, one can then use YouTube's 'Annotation' feature to link to another video. The one drawback is that you can only link to other YouTube videos. This procedure adds additional steps once you've created your screencast and will most likely result in a screencast that is less attractive than one in which the video is embedded. So, in terms of open markets in education, a password protected site is, for now, the better solution. Models for this solution include open education websites like Khan Academy,[20] where users have to create a login and password in order to access the material.

Further reading on how to use copyrighted material and its rights in open education can be found in the free PDF *Code of Best Practices in Fair Use for Media Literacy Education*.[21]

Conclusion

As technology continues to advance in ways that might enable innovative educators to enhance teaching, it is always important to bear in mind that pedagogy should take precedence over technology. Technology is cool and oftentimes fun, but there must be a reason for incorporating it into one's course to enhance student learning. If that turns out to be the case, as screencasting has for my students, it then becomes a matter of ascertaining the best way of making that material available. Some solutions are already available, such as hosting a course site on Weebly for Education,[22] which has video hosting capabilities as well as password protection. Such distribution is not entirely open, but, hopefully the concept of open education will evolve to the point where distribution of creative endeavors can more easily be accommodated.

Notes

[1] At School, n.d.
[2] Krause, 2000.
[3] Petroski, 2005.
[4] Petroski, 2005.
[5] Mann & Robinson, 2009.
[6] e.g., Schmaltz & Enström, 2014.
[7] Examples, n.d.
[8] Daniel & Woody, 2010.
[9] ScreenFlow, n.d.
[10] RecordIT, n.d.
[11] OpenShot Video, n.d.
[12] Psychological Science, n.d.
[13] iCube, n.d.
[14] e.g., Young, 2011; Ledonne, 2014.
[15] Wistia, n.d.
[16] Wikimedia Commons, n.d.
[17] Wikimedia Commons Own Work Rules, n.d.
[18] Flickr, n.d.
[19] British Library, n.d.
[20] Khan Academy, n.d.
[21] Media Education Lab, n.d.
[22] Weebly for Education, n.d.

References

At School. (n.d.). Available at https://commons.wikimedia.org/wiki/File:France_in_XXI_Century._School.jpg

British Library. (n.d.). Available at http://www.bl.uk/aboutus/terms/copyright/

Daniel, D. B., & Woody, W. D. (2010). They hear, but do not listen: Retention for podcasted material in a classroom context. *Teaching of Psychology, 37,* 199–203.

Examples. (n.d.). Available at http://tedxuconn.com/david-miller/ and at https://www.youtube.com/watch?v=dIMiba75LBA

Flickr. (n.d.). Available at https://www.flickr.com/creativecommons/

iCube. (n.d.). Available at http://icube.uconn.edu/InstructorAdvice.mp4

Khan Academy. (n.d.). Available at https://www.khanacademy.org/

Krause, S. D. (2000). Among the greatest benefactors of mankind: What the success of chalkboards tells us about the future of computers in the classroom. *The Journal of the Midwest Modern Language Association, 33,* 6–16.

Ledonne, D. (2014). Multimedia assignments: Not just for film majors anymore. *The Chronicle of Higher Education: The Digital Campus,* B30–B31.

Mann, S., & Robinson, A. (2009). Boredom in the lecture theater: An investigation into the contributors, moderators and outcomes of boredom amongst university students. *British Educational Research Journal, 35,* 243–258.

Media Education Lab. (n.d.). Available at http://mediaeducationlab.com/sites/mediaeducationlab.com/files/CodeofBestPracticesinFairUse.pdf

OpenShot Video. (n.d.). Available at http://www.openshotvideo.com/2016/01/openshot-20-beta-released.html

Petroski, H. (2005). Next slide, please. *American Scientist, 93,* 400–403.

Psychological Science. (n.d.). Available at https://itunes.apple.com/us/podcast/psychological-science/id1016929637

RecordIt. (n.d.). Available at http://recordit.co

Schmaltz, R. M., & Enström, R. (2014). Death to weak PowerPoint: Strategies to create effective visual presentations. *Frontiers in Psychology, 5.* DOI: 10.3389/fpsyg.2014.01138

ScreenFlow. (n.d.). Available at http://telestream.net/screenflow/overview.htm

Weebly for Education. (n.d.). Available at http://education.weebly.com

Wikimedia Commons. (n.d.). Available at https://commons.wikimedia.org/wiki/Main_Page

Wikimedia Commons Own Work Rules. (n.d.). Available at https://commons.wikimedia.org/wiki/Commons:Own_work

Wistia. (n.d.). Available at http://wistia.com

Young, J. R. (2011). Across more classes, videos make the grade. *The Chronicle of Higher Education.* Retrieved from http://chronicle.com/article/Across-More-Classes-Videos/127422/

Librarians in the Pursuit of Open Practices

Quill West

Pierce College, CWest@pierce.ctc.edu

Editors' Commentary

Any serious book on open would be incomplete without the inclusion of the librarian point of view. Libraries have long been repositories of learning that are strongly aligned with the open philosophy. In this chapter, author Quill West speaks to the lofty aspirations of open education. Refreshingly, her commentary is as pragmatic as it is idealized. She offers practical insights into the various ways that faculty stumble over open. West is a realist in her admission that not all open materials will satisfy every course learning outcome. Deficits, where they exist, can often be addressed but many faculty are not certain exactly how to do this. She views openness—ranging from publishing open materials to giving attribution to others—as a competency that can be learned.

Do you remember that day, or hour, or moment in your educational career when you first turned to a librarian for help? Librarians have helped us find materials when research got difficult. They helped us recreate our reference lists when we lost track of citations. They cared for the spaces, books, databases, and information tools that gave us refuge when studying. As we grow in the academic professions, librarians continue to offer those valuable information skills in planning classes, researching professional projects, and supporting student learning initiatives. Basically, librarians are all-around, interdisciplinary, learning-centered professionals who encourage the best teaching and learning

How to cite this book chapter:
West, Q. 2017. Librarians in the Pursuit of Open Practices. In: Jhangiani, R S and Biswas-Diener, R. (eds.) *Open: The Philosophy and Practices that are Revolutionizing Education and Science.* Pp. 139–146. London: Ubiquity Press. DOI: https://doi.org/10.5334/bbc.k. License: CC-BY 4.0

by providing services meant to challenge people to interact with information critically and responsibly.

This piece runs the risk of being an ode to the virtue of librarianship, but I believe that to understand the value of librarians in your open practice, you must see the value of librarians in the academy. In fact, as an open librarian I find it difficult to tease out the differences between my work in open education and the professional practice of librarianship. At my core, though, I want to express one simple truth:

> Your institutional library team should be one of your first calls when you decide to explore open education for the first time, just as it was natural to turn to your librarian the first time that you couldn't find a resource that you needed for a particular assignment.

Open Practice – Open Values

Open education is a philosophy that seeks to create equal access to education through lowering costs and increasing relevance of materials used in teaching and learning. While that is a generous and timely goal, the actual professional practice of adopting and adapting open materials, indeed even creating open materials, is part of a larger practice where an individual engages in open as a way of designing, teaching, and distributing a course. The ideals promised in open education are achieved when we practice the application of six open habits until they are a part of the way we live our professional lives. While open librarians often encourage teachers and learners to respect many of the principles of open education, sometimes it is hard to know proven practices that lead to easier course creation. We achieve openness by exploring and encouraging the six habits of open practice: sharing, early drafting, supportive feedback, studying licenses, giving credit, and putting students at the center.

Sharing

Sharing is an art that depends on the ability to know when to share something, and whether it is useful to the people we are sharing with. Moreover, sharing can be vulnerable and uncomfortable for many reasons.

First, most academics spend years learning the value of intellectual property and it is difficult to 'give it away'. However, in my own practice I have often reminded myself that sharing isn't giving something away. By its very nature knowledge cannot be given away, because the person giving retains the knowledge even as she passes it to someone else. Indeed, sharing our knowledge is a central piece of what instructors do every day. However, it oversimplifies the issue to ignore that intellectual property does have economic value in some instances. When I began sharing my works with open licenses, I balanced the potential future value of my creative work against the immediate, and definite, value of a community of people to help improve and grow the materials that I share, and I decided that sharing was of greater value to me. I have asked colleagues who are concerned over loss of possible future royalties to consider how much effort they will expend to turn their current work into a resource that a publishing company might be interested in, and how few textbook authors actually make significant royalties on their textbooks, and then I ask them to weigh those ideas against the power of a community of users and adapters. Usually the community ends up having more significance in the current life of the creator.

Another concern of most people considering sharing their openly licensed work is that adapters could use the work in a context which isn't intended, or that the adaptation could get information wrong. It's true once a work has been shared the original creator has almost no control over what is done with it. Another teacher could remix your work in a way that you would disagree with. An adapter could misrepresent you or get your work wrong. Reputation is one of the major contributing factors to hiring and promotion, particularly in academic settings, so the risk of having one's work altered without permission is a serious one. However, the other side of this argument is that you keep your work pristine, and safely in your own computer only to be shared with students, and no one ever sees your work. I try to help my colleagues see that rules surrounding adaptation and attribution of openly licensed work asks the adaptor to describe how the work was altered. Also, an adaptation always includes a link back to the original work. I also try to remember that even if something is published with the standard copyright protections, there is no protection from being misquoted or taken out of context. Once, I tried to talk a faculty member into sharing an extensive collection of handouts that she had prepared. She was worried that people would revise her work, and make it less effective. After several weeks of discussing her worries over sharing her work, and having it changed, she finally decided to openly license the handouts. I asked her what changed her mind, and she told me that it was actually a student. The student annotated and wrote notes all over the handout, and then passed it to a fellow student. She realized that the student, a complete newbie to her field, had just improved the original design by making it more accessible to his classmate. She wondered what useful adaptations others teachers of her discipline might make to improve the final document.

Another difficulty that I have encountered when encouraging others to share is predicting what others will want. Teachers, in particular, often say a variation of the following to me: 'No one could use what I do in my class, because I do it my way.' How do you know when you should go to the trouble of making something available, especially when you don't know how useful it will be to other people? First, I always ask people, 'How do you know no one else can use it? Maybe you're revolutionizing teaching this topic and no one else will ever know because you aren't sharing it.' However, that answer might be overly optimistic.

Instead, I try to think in terms of asking for an expert to help decide what might help others. Here again, librarians are useful people. Librarians most often work with your colleagues across multiple disciplines, and they most definitely work with students from across your institution. They can help you to see if your material might be useful to others.

Finally, sharing is made more difficult by not knowing how to get your work into a sharing space. The logistics of moving a creative work from your computer, course management system, notebook, canvas, or video camera to a space where others can find and use it is one of the more frustrating issues when you seek to share. I have faced the same challenges in my career. One of the biggest talents a librarian can bring to your sharing is a professional interest in curating and sharing knowledge. As a librarian I have had to invent low and no-cost strategies for sharing my own as well as colleagues' works in front of a worldwide stage. At larger institutions the advent of institutional repositories and digital scholarship departments means that space and encouragement for sharing are more prevalent and respected. Check with your library, because chances are someone might be able to help you to share your work more widely.

Early Drafting

One of the biggest gifts that open can bring to your educational practice is the permission to stop being a perfectionist. Educators put a lot of pressure on ourselves to create perfect works. We don't submit journal articles to peer review until they have been internally vetted, and we never expect to be published if our work doesn't meet a gold standard. We rely on extensive editorial processes to help us pick over our creative works so that they will be ready for our colleagues to view. The review process is necessary and important in academic scholarship. In fact, the review process is so necessary in higher education that one of the common criticisms of early open materials is that they didn't include that process. However, I would argue that this is a value of the open education field because instead of limiting the process to a select few people who will be responsible for vetting a work that might be intended for an entire field, open education resources are vetted by everyone in the community. Some open materials are reviewed, revised, reworked, remixed, and continually updated by an entire community of people who are passionate about maintaining the resource.

It's important to break here to note that some open projects, such as OpenStaxCollege, BC Open Textbook Project, and NOBA have implemented extensive review and editorial processes. These projects should be celebrated, and relied on when possible. However, most individuals seeking to participate in open resources by contributing lessons, handouts, lectures, notebooks, visuals, tutorials, and writings won't find that kind of support easily. I will note however that librarians at your institution might be a useful place to find out if there are extensive projects like the ones described that your work could fit into. A growing number of librarians in the world are following open creation and adoption projects.

As a creator the gift that open education brings to you is simply: it's okay to not be perfect. This isn't to say that you shouldn't proofread, or ask local colleagues to read and give input on your open materials before you release them. It does mean, however, that you can start to share things earlier and more often.

I see faculty embodying this ideal of early drafts on a regular basis. A great example is the work that faculty at Pierce College at Joint Base Lewis-McChord (JBLM; a military base) has begun sharing recently. They used the Lumen Learning platform to revise, remix, and write a series of resources for Introduction to Nutrition. As a strategy in the design phase of that work, the faculty team completed the textbook enough to teach the first quarter with it, and then they released the book through librarians, instructional designers, and open project managers. The goal was to ask for the community for suggested revisions and additions. Eventually the book will be revised, and hopefully the community of teachers of Nutrition will help with revisions.

Supportive Feedback

If one of the habits of open educators is to release work before it goes through an extensive review process, then it falls to the community using the work to give respectful, honest, supportive feedback that will move the general work forward. As educators adopt open materials it is increasingly helpful to let the creator know what materials you are adopting, how you are using those resources, and any remixing or revision you might have done with the work. I'm fond of telling students and faculty that we don't know if a work needs to be better if no one tells us.

Studying Licenses

You don't have to be a copyright expert to apply open licensing to your own work. You do, however, need ensure that you are using other peoples' works ethically and legally. An essential part of your open practice should be examining works for licensing and usage rights. While open licenses are designed for ease of interpretation, the implications of mixing and matching licenses can be frustrating.

Most institutions have librarians who have studied how copyright and open licensing impacts their institutions. Cable Green's chapter in this book is a helpful description of open licenses and their use, which you might consider sharing with your library team. If your local library team doesn't have ready expertise in open licensing, they probably have a network of people who can help them to address your question.

Giving Credit

Crediting original authors of ideas is not a new concept in higher education. In fact in libraries all over the world there are students creating American Psychological Association (APA) citations right now. However, it isn't a normal practice to include attributions on all of our slides or in all of our writings. We are used to crediting ideas, but not necessarily artifacts of those ideas. There are a series of best practices for writing and including attributions on openly licensed works.

I encourage faculty to develop a practice of always including an attribution on every creative work adapted. More than once in my academic career I have been tasked with going back to add attributions to works where they are missing. My least favorite conversation to have with a faculty adopter of open materials is that material has to be omitted or substituted because the attribution cannot be verified. That is why it is incredibly important to start with the right attributions to begin with.

Building a consistency around attributions and recording rights is a habit that all users of open materials should grow into. Certifying in-house created and remixed open materials is also an important part of the attribution process. Librarians can be helpful allies in reviewing and verifying attributions and ethical use of openly licensed materials.

Putting Students at the Center

Students, their learning, and their access to creative, energizing, and engaging educational experiences is at the center of most open education projects. Teaching practices should always consider how students interact with materials, lessons, assignments, and one another. When reviewing open materials, I have always tried to picture the students who will be asked to use the resources. I try to picture myself as a reference librarian tasked with explaining the resource to a student. What kinds of questions will the student arrive at? How will the student internalize the concepts presented? Will the student find rewarding experiences in doing the assignments that surround the open material? These questions guide the process of evaluating and enhancing open materials to fit the instructor's teaching practice.

Librarians are in unique places at our institutions, because they often interact with students across disciplines. In my experience, librarians are usually the best people to help clarify assignment prompts because they have so much face-time with students of varying abilities and academic experiences. While instructors spend several hours over the course of a quarter with a group of students, librarians often spend one on one time helping students from across the institution to define and articulate information needs, so that students can successfully complete course assignments.

Another level of putting students at the center of open education initiatives is inviting student voice to the planning and implementation of overall projects. Librarians can assist in this conversation, because students often see libraries as safe places to share opinions and ideas. Librarians and libraries have professional values of service and support of our patrons. These incredibly central professional ethics have built a decades-old culture of safe spaces, which can be leveraged carefully to encourage students to share ideas for bettering their experiences. At some institutions the average student voice is illusive, no matter how hard we try to capture it. In some cases, librarians can help to solicit the quieter student voices. I have seen students tell librarians about personal financial issues, challenges in access to basic services, and issues regarding college readiness. Admittedly, student service offices can serve the same purpose. However, librarians are in a position to bridge both student service and academic approaches in outreach to students. When possible and where appropriate, librarians should be included in planning to invite the student voice into open education.

Open Education and Information Competency

One of the most exciting elements of open practice is the opportunity to improve teaching and learning. Adopting open materials can include a process of examining and improving pedagogy. Early in my open education practice I realized that sometimes open materials will fall short in terms of total coverage of course outcomes. This posed little threat to me, because as a librarian I knew that there were other ways to encourage students and instructors to fill in content. Information competencies, and assignments designed to build both content information and information skillsets, are a way to further student learning and meet course outcomes. An information competency assignment might ask students to research and summarize major concepts of course outcomes. By conducting research on their own, students are more likely to remember course content. Also, they are encouraged to see that knowledge building and information creation is a process that includes an interaction between knowledge seekers and information sources. I'm hardly the only librarian to arrive at the conclusion that open education is a way to encourage students to interact with a cycle of information that includes the ongoing process of creating, evaluating, and incorporating new knowledge into existing sources. In fact my voice is one of many advising open practitioners to include information competency assignments as part of growing pedagogical approaches to support open education.

Conclusion

Information and using it to grow knowledge and wisdom is a central part of what being a student is all about. Open practices reinforce this idea by making the information creation process a little more transparent to students. Instead of being faced with a pre-packaged, comprehensive collection of readings that students often interpret as the 'learning' of a course, open materials ask the students to invest in study practices that include active decision-making about learning materials. In this way open practices align more completely with missions of libraries because it is a central value of our profession that all people need to develop information competency. Also of central value to our profession is the belief that sharing resources – books, computers, space – improves the larger community. Open education is a growing educational movement where these two values meet. By sharing our expertise in curating resources, building information competency, serving students and institutions, and in moving across disciplinary silos, librarians can help our institutions to embrace change that will open access for many of our students.

Most academics love librarians because librarians are dedicated to stripping the confusion and mystery out of finding and using information. Librarians make being a student easier, which helps students to meet their goals. However, the truly savvy professor knows that their institutional library team also makes teaching easier.

A Library Viewpoint: Exploring Open Educational Practices

Anita Walz

Virginia Polytechnic Institute and State University, arwalz@exchange.vt.edu

Editors' Commentary

There are a wide range of faculty responses to open; everything from curiosity to resistance. Some of these reactions are due to professional tradition, guild thinking, and lack of awareness about open. Academic libraries offer a unique context for exploring open. They do not fall prey to disciplinary concerns and, as such, are hotbeds of collaboration and innovation. It is no wonder that libraries are a natural home for open educational philosophy. In this chapter, author Anita Waltz offers her personal insights into the current state of open education. She candidly shares her thoughts on faculty adoption of OERs, the current cost of learning resources, and the promise of open pedagogy.

In the Spring of 2014 I began trying to enlist faculty for an Open Education Week panel discussion – our first at the Virginia Tech's University Libraries. I talked with seventeen faculty members regarding their thoughts on textbook adoption and selection of learning materials with the hope that faculty members would freely and publically share their thoughts about selection or design of learning materials. One said yes right away. Some never replied. Some were, themselves, textbook authors and told me of their experiences writing, designing, and formatting their textbook and the resulting miniscule royalties – which they did not want to lose. Several had adopted 'custom textbooks' but did not want to talk about this publically. Several would have been interesting

How to cite this book chapter:
Walz, A. 2017. A Library Viewpoint: Exploring Open Educational Practices. In: Jhangiani, R S and Biswas-Diener, R. (eds.) *Open: The Philosophy and Practices that are Revolutionizing Education and Science.* Pp. 147–162. London: Ubiquity Press. DOI: https://doi.org/10.5334/bbc.l. License: CC-BY 4.0

contributors to a panel discussion but suggested I talk with their colleague instead. A panel with one panelist is not a panel. Hours of interesting conversations – yet none of these knowledgeable people were interested in speaking publically on this topic. I was bewildered and frustrated. I concluded that discussing problems regarding teaching and learning resources is somehow uncomfortable or otherwise not rewarding. And so, this is how I started my exploration of open educational practices at Virginia Tech.

Open Education is a philosophy which prioritizes identification and removal of many types of barriers to education and learning. In its very broadest sense it may mean high quality educational opportunities freely accessible by anyone regardless of location, enrollment status, or ability to pay. Many people ask if 'open education' is synonymous with free tuition or with the removal of admissions barriers so that anyone can learn. This is the case with MOOCs (Massively Open Online Courses), though very few are accredited. In the case of The Open University (UK) payment of tuition is required but no admission or entrance exam requirements exist for undergraduate level courses. In higher education contexts, at least in North America, free tuition and no admissions requirements are typically not what open education means. In North American contexts, open education typically aims to reduce access and cost barriers to learning materials, prioritizes student engagement, agency, access to information and ideas, and relevance of course work to the real world. Overall, open education practices seek to improve educational quality and access.

I've been a librarian for over fourteen years in many different settings, mostly in international, and subject-specific 'special' libraries, often with a very specialized role. In terms of consultation services, collections, research, and teaching responsibilities my past roles felt similar to academic libraries. Six months into my new position at Virginia Tech, I realized that academic libraries – especially in research universities with tenure tracks – are a completely different animal than I had encountered before. Unlike other types of libraries, librarians in academic librarians enjoy full participation in institutional governance, enabling access by librarians to the interests and concerns of non-library faculty – and deep information regarding how a University actually works. Due to their tenure path, academic librarians may have more agency than other types of librarians. This allows academic librarians to interpret their role in order to fit program and institutional needs. Further, the culture shift in some academic libraries towards innovation and collaboration lend academic libraries a broad landscape for creative opportunities in scholarship, teaching and service. These realities require deep understanding by academic libraries of teaching and research faculty roles, values, pressures, and processes, each of which are relevant to my journey. As Virginia Tech's Open Education, Copyright and Scholarly Communications librarian I'm tasked with exploring potentials for and disseminating information, resources and support regarding Open Education at Virginia Tech and beyond. This brief narrative summarizes my experiences learning about and advocating for open education efforts over the last several years.

Open Education Immersion: Sink or Swim?

My first introduction to open education was not that long ago. I attended the OpenVA conference in the Fall of 2013. The two-day event, including the 'Minding the Future' preconference was packed full of faculty excited about teaching, technology, and reducing access and cost barriers to student learning materials. It was fun. It was mind blowing. Some of it seemed crazy. Some of it was crazy – such as an online class without an instructor! Or asking students to create digital identities on their own domain! I now realize it was a good kind of crazy; nevertheless one I was not ready for at the time. At that time, the reduction in learning material costs of Tidewater's Zero Textbook Degree was the only thing I could wrap my brain around; the ideas of open pedagogy were a bit too much for me as part of my first introduction. But clearly other people were excited about these ideas and I have since worked to better understand and appreciate these aspects of open education.

Attending the OpenVA conference created a strange chasm between what I heard there and the realities of the institution to which I was beginning to acclimate. In contrast to the large number of faculty excited about open educational practices at Open VA, I could not find any faculty members at my institution talking about open educational resources or open pedagogical practices. It was challenging enough for me to understand and describe what I was looking for. Are any faculty members adopting these methods of authentic assessment or asking students to create things that are viewable or have value outside the classroom? Who are they? How do I find them? Are any faculty authoring or using open educational resources? I still hadn't *seen* any of these seemingly-mythical open educational resources people were talking about. What are these? Are they books or something else? Are they bit of this and that cobbled together? Could that be any good? Why would someone want to give away something they spent a lot of time and effort to create? Some answers came through philosophical conversations with my new supervisor and reading Lawrence Lessig's *Remix*, Kevin Kelly's *New Rules for the New Economy*, and other books and articles. Some came when I realized I could integrate a discussion of Creative Commons into Copyright education sessions. Some answers came when I was introduced to OpenStax College, a project of Rice University, which creates full, complete textbooks for high enrollment, intro level courses. They create textbooks then put the most open (i.e., least restrictive) Creative Commons license (CC BY) on them and post them to the web in multiple formats (see Chapter 17 in this volume). I met faculty who authored them (and told me they were paid), and faculty who were using them in courses. Wow! This is great. These would be really helpful for students – if they are as good as those who are currently using them say that they are. Some answers came thru joining SPARC's LibOER Listserv and being graciously granted permission to lurk on the Community College Consortium's OER Listserv and attend their free webinars.

The openly licensed books were amazing. Full color, real, around US$40 in print and free in PDF and other electronic formats, as well as extensively peer reviewed. My faculty will snap these up – I thought. I introduced a few faculty to the books – and waited and asked. Nothing happened. Why aren't people clamoring over no-cost learning resources? Is it me? Is it them? It is because I'm new and they don't know me? I wasn't sure. I thought these would be welcome. This would be good for students and student learning. Why aren't my faculty interested? I certainly did not expect faculty disinterest regarding open education resources. Though given my recent move to academic libraries, I also knew that I had a lot to learn regarding the nature of faculty work.

Information about faculty concerns came from unexpected places. I was invited to speak about my first year working as a librarian on Open Education, ironically at the OpenVA 2014 conference. I felt very isolated in my exploration of open education yet persisted believing it to be of value; I knew what I thought was just a little bit and was willing to share it. I was thrilled to find out that were three or four other librarians at OpenVA 2014. More importantly, I found some clues regarding faculty concerns at OpenVA 2014. One faculty member gave a presentation on how she got started in open education. She reported covertly 'becoming a little OER-ish' as she stealthily explored and implemented open resources and open practices in her courses. The fact that a faculty member would be nervous about how her colleagues would respond was new to me; It finally dawned on me – maybe I couldn't find faculty at my institution doing the same, because they didn't want or weren't ready to be found. Again, this is not what I expected but meeting this faculty member was an important clue in my search.

[Mis]understanding Faculty

I decided that I would be transparent about my challenges in obtaining faculty interest. My presentation at the Open Education 2014 conference was titled 'What Faculty are Actually Doing.' In actuality, my presentation was about dealing with failure, encountering resistance, adapting a learning posture, and choosing to persist. I had made several false assumptions: I assumed that faculty would be eager to talk about and would clamor to adopt OER; I had also assumed that I understood campus culture and faculty needs. Both of these assumptions were erroneous and my seventeen conversations with faculty members as I was planning the Spring 2014 panel discussion still resonated. While this list is by no means exhaustive, there seem to be three main themes emerging that described reasons for faculty indifference and even resistance:

First, low faculty awareness of open educational resources is well documented; skepticism regarding openly licensed resources is also a common first response. 'Free and high quality' seems to invite a good deal of healthy skepticism. I resisted initially too: are OER 'real' learning resources? Why would someone freely give something away if they could make money off of it? What's the catch? In general, the idea of openly licensed content *is* a bit shocking. We tend to think that things we get 'for free' are junk. It's fairly revolutionary that one can now find peer-reviewed, lengthy, legally posted, complete, and because of CC BY licensing, editable textbooks and other types of learning materials. (Apparently over 2,000 people on Twitter thought this was exemplary as well.) Nearly every week I have the privilege to tell faculty or students about OER, open licensing or Creative Commons, and how they can use or author such resources. Sometimes I think we won't get past this introduction stage, but I'm committed to explore this topic with people for whom it is brand new. It is always 'day one' for someone to learn about OER. It's a lot to absorb.

The second reason for indifference is risk-aversion. Asking people to change how they do something is a big deal – not just in terms of time or effort – we'll get to that later, but because it's uncharted territory. It's new. It can be scary. The consequences are largely unknown. On more than one occasion a faculty member exploring or using open educational resources has told me they don't want 'to go public' or 'Please don't tell anyone I'm exploring open textbooks.' Some faculty are risk-averse presumably out of concern that adopting or authoring open educational practices will reflect poorly on their reputation or career. One anonymous faculty member reported that his colleagues would look down on him if he chose an open text instead of one from a prestigious publisher (unpublished 2015 survey). While I'm not privy to all faculty members' pressures, I do know that reputation, peer and administrative relationships, and tenure pressures weigh heavily on faculty. Faculty do not have uniform knowledge or expertise regarding learning resources in their disciplines. Some may never have taken time to review an open textbook. But perception of quality or lack thereof can be very influential even if it is unfounded. An even more risk adverse group of faculty are faculty authors of commercial textbooks. I don't go out of my way to talk about OER with authors of commercial textbooks as I find these conversations to be very awkward! Authors rightly take pride in their hard work and investment as an author. They often enjoy the collaborative support of editors and publishers who recognize their accomplishments and value their work. While only the top few authors in a field receive substantive royalties from commercial textbooks, and most authors don't write textbooks for the income it can be very difficult for an established author to change course. Once a revenue stream is established it certainly is hard to shut it off! Then, there is the issue of the author's agreement with their publisher. Some publication contracts I've seen assign the publisher rights for every future edition of the book and waive rights to publish a similar work elsewhere. Depending on the agreement and short of author rights reversion, there are not many options for creating OER unless the created work is on a different topic than the textbook. When the inevitable conversation occurs, the first thing authors usually mention that open textbooks might undercut their potential profits – and then they take a defensive stance. I'd love to find better ways to work with these hardworking, potential authors of open educational resources.

The third reason relates to inertia and investment, or the lack thereof. When the status quo seems to be working (for faculty) authoring materials or integrating already-created openly licensed content into an existing or new course requires one to prioritize making a change. Exploring or adopting or openly licensed works, reviewing open textbooks, and spending effort required to make courses adjustments may be difficult when the status quo seems to be working. While some faculty chose to pilot an open text as an option for their course as a way to test the waters, I've also had multiple faculty tell me 'what we're using works for us.' I'm grateful that something is working for them, but I'm also very hesitant especially when there are many openly licensed options for certain courses, and when I know that many students are very frustrated by the cost of textbooks, homework software codes, and are either sharing books, going without, or taking on extra hours of work to afford it all. That said, faculty academic freedom clearly puts curriculum materials and teaching methods in the purview of the faculty member(s) teaching the course or of a departmental committee. My role regarding this area is to inform faculty of options available to them and offer review, authoring, and course-adjustment opportunities. Some of this reticence to make changes may be related to institutional type, tenure-status, and perceptions of whether or not the cost of learning materials are really that severe for students. Within a research institution promotion and tenure-related requirements often take precedence over other types of activities. There is never enough time and the many responsibilities of being a faculty member can be overwhelming. Even at institutions that do not emphasize research and publication as much as research institutions, there is always competition for time and effort.

Then, there are faculty who wish to explore and adopt open educational resources for which only limited content exists. Faculty teaching specialized topics or upper divisional courses often face an insufficient amount of openly-licensed content. This is logically the time to develop new openly licensed content as time allows, which it does not always allow. Other options include assigning an older edition, finding a lower-cost work, conducting a Fair Use analysis for portions of copyrighted works, or finding an interim solution of using no-additional-cost library subscribed resources. Neither of these are perfect solutions. The library solution merely shifts costs from the student to the institution's library, which some but not all libraries can or want to handle. For those which can afford buying multiuser site licenses at the moment, this is probably not sustainable in the long run. Further, multiuser site licenses are not available from all publishers, and students lose access to important works after they graduate. As an interim solution, the mix of openly licensed and commercial licensed works is somewhat inevitable due to a lack of content, but should ideally be considered an interim solution. In summary, resistance and indifference come down to effort. It's easier *not* to take on the effort to explore and consider change.

Librarians in Open Education

In addition to raising awareness, one of my main roles is to find ways to support faculty exploring and adopting open educational practices, so understanding these points of indifference and resistance are very important. Librarian roles around Open Education are still emergent. Not many librarians have the luxury of devoting more than a little of their time to open education related work. Most librarians, including myself have multiple responsibilities which compete for time and attention. Contrary to popular impressions, librarians don't have time for leisurely reading during the work day and some of us don't touch a physical book for weeks on end. We are involved in teaching, subject liaison activities, program and instructional design, inventing, mastering, and leveraging new technology. Many of us are involved in web development, building systems and processes, collaborative partnership building, research, consulting, curating data and collections, purchasing, conducting research, writing, giving presentations, and on it goes. Our real and existing roles as problem solvers, user advocates, teachers, and those who seek to make sense of information and research trends and tools give us a tremendous foundation on which to build and innovate.

To the open education movement, we bring a tremendous wealth of knowledge and expertise in copyright and licensing, inquiry-based learning, user advocacy, systems thinking, project management abilities, and expertise in teaching. Many of us work hard to ensure that we have a place at the table when learning resources, educational technology, rights, and pedagogy are discussed. And many of us lead institutional initiatives in these areas. Depending on our main roles and the needs of our institution we may implement and connect open educational practices very differently. There is no single model for librarian involvement in open education; I think this is a good thing. I have seen extraordinary, inspiring librarians working in open education in their areas of strength in instructional design – conducting what we call 'reference interviews' for curriculum topics and training librarian leaders. An increasing number of librarians lead, design, and manage faculty development programs with OER author and adopter incentive grants and a huge impact. Many teach and consult on Copyright and Creative Commons licensing. Some librarians engaged in open education lead multi-week faculty development sessions on open education. I see librarians building and running project networks to support faculty in state-wide multi-institutional initiatives, and librarians building and managing impressive national networks to support and train librarians and others regarding this type of work. Librarians are increasingly building collaborative relationships with faculty, instructional designers, academic support personnel, concerned students, and administrators. I am extremely proud of my librarian colleagues' leadership, creativity, and achievements in this field.

My own work has focused on identifying policy and institutional barriers to open licensing and methods thru which faculty may openly license their work if they choose to do so. I've met with our University Legal Counsel and Intellectual Property Committee. I've worked as a Co-Principal Investigator on creating openly licensed works for teaching. I've been involved in the arduous process of revising an open textbook. I support other units and groups who want to integrate information about open licensing, OER and Creative Commons into their courses, and teach and consult with faculty involved in course redesign. I work with student groups exploring these issues. I conduct research to better understand our campus context and suggest implementation options that may fit better than some others. I serve on a state-level committee and have prioritized making outside speaker events and open education week events open to the public, this year taking the step to live stream public programs.

A large part of my role beyond other job duties is advocacy and raising awareness around open education. This includes events showcasing the work of Virginia Tech and nearby faculty and students. It also includes research.

Students' Learning Resource Buying Patterns

I knew from past work with the Student Government Association that students had a lot to say about textbook costs, and I wanted to get beyond anecdotes. At Virginia Tech the Admissions office indicates that students should expect to pay US$1,000 per year on textbooks and supplies. I believe this to be somewhat of a conservative estimate given that the College Board's estimate for 2015–2016 recommends that students at 4-year universities expect to pay US$1,298 on

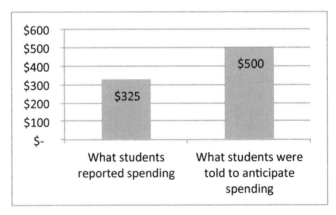

Chart. 1: Reported and Anticipated Student Spending, Virginia Tech Spring Semester 2016

Source: Walz, A., Spring 2016 Virginia Tech Student Survey (n=312) (Unpublished) and Virginia Tech, Department of Admissions.

books and supplies. So, during weeks 3–7 of Spring 2016 I invited a representative sample of 3,000 Virginia Tech undergraduate and graduate students to take an anonymous survey regarding learning resources. While this is just one data point, I think the illustration is important; what I found surprised even me. I thought that students would report spending only slightly less than US$500 per semester on learning resources; students reported spending an average of US$325 on textbooks, required learning software, and supplies in Spring 2016.

This US$325 is 35% less than what admissions office indicated they should budget. While this is surprising, it also affirms national survey findings in which students increasingly view textbooks as optional. This tells me that students are even more price sensitive than I suspected.

Disclosure of costs of required learning resources are mandated by the Higher Education Authorization Act (2008) at the point of course registration, and institutions Admissions offices disclose anticipated costs. Somehow the costs of learning resources are still overlooked or come as a surprise. Perhaps this is because of they are a much smaller cost, at less than 10% the cost of tuition at Virginia Tech, though a much higher percentage of overall cost at Community Colleges. Perhaps students or parents are not budgeting to pay for learning resources. Perhaps the responsibility and cost of learning resources are passed directly on to students for them to deal with after their first semester or first year. While proportionally smaller than tuition in cost, learning resources (or the lack there of) can have an enormous impact on one's academic achievements. In courses where textbooks are strongly recommended but not required, faculty have mentioned that students without access to the textbook consistently earn lower grades. Is it any surprise? There is another area of impact: for first generation, low income, and the first child in a family to enroll in college these costs may come as an unwelcome surprise and create a barrier to academic achievement. I suspect that these costs may disproportionately affect the academic achievement of an institution's most vulnerable students. It may also interfere with institutional efforts to increase socio-economic diversity on campus.

I have initially been very hesitant to talk about cost of learning resources as a motivator for openness. The details were very vague; economic issues of students felt too personal and maybe too political. Who wants a rant to listen to another rant about cost? In the context of higher education though, cost is a central issue for nearly everyone: University budgets of public institutions are a topic both in the State legislature, University and Departmental level. Administrators concern themselves with how to strategically budget in for current needs and strategically invest in facilities, programs, and personnel for the future good of the University. Families with students, students, and students with families face a financial calculus of their own. A recent spate of books on the topic including: *Why Does College Cost So Much?*, *Is College Worth It?*, and other titles aim to dissect the college cost situation and question at what point paying for a college education is still a wise use of resources. Neither the cost of tuition nor the cost of learning materials are like they used to be. If

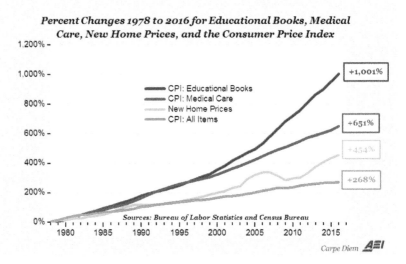

Chart. 2: Percentage Change from 1978 to 2016 for Educational Books, Medical Care, New Home Prices, and the Consumer Price Index (CPI).

© 2017. Mark Perry, Professor of Economics, University of Michigan-Flint and Scholar, American Enterprise Institute. All Rights Reserved. Used with permission.

you attended college between 1980 and 1990, your textbooks cost roughly one-quarter to one-third of what textbooks cost today. The cost of college textbooks has increased nearly four times the rate of inflation since 1978.

Many students are increasing required to purchase homework software access codes. Students in courses where an access code is required cannot get a grade without purchasing an access code. And, unlike used textbooks, used codes are not transferable or reusable. Software codes, depending on the cost, can also present barriers for students.

While software codes are required, faculty mention that 'Can we share the textbook?' is the first question students ask about the textbook, ahead of 'Do we really need the textbook?' Students respond to textbook costs in many different ways. As Dave Ernst from the Open Textbook Network taught me, 'they either have the money, borrow the money, or earn the money.' Some will obtain an older edition, in hopes that it will be similar enough. Some will share books. Some will expand their paid working hours or get a second job. Some will wait until they are behind in the course to buy or rent the material. Others, as exemplified by the survey data above choose to go without. Even though learning material costs are a smaller cost than tuition, these proportionally smaller costs are often overlooked and can have a disproportionate impact on academic achievement.

Faculty Awareness and Decisions Matter

Unlike tuition and room and board, faculty decisions have a direct impact on these barriers. Some faculty members are attuned to student costs; some are not. Even though I thought I was in tune with student price sensitivity, I was not nearly enough aware. Students are even more price sensitive than I thought.

There are many things faculty members can do to aid students with regard to cost, many of which have been alluded to earlier or in other chapters of this book. Getting buy in first from one's department and colleagues, especially if teaching in a sequence of courses and recruiting assistance from a knowledge-able colleague, librarian, or instructional designer are excellent steps. Including cost as one criterion for learning resource selection by committees or your own course is a good start. Ask students what they consider to be 'expensive.' Giving students more choice is another; some faculty require students to obtain a textbook and allow students to choose one of three different options – one being an openly licensed, no cost option. This is a low-barrier way to pilot an open textbook or other open learning materials. While not randomized, this may also lead to interesting research regarding student choices and academic achievement. Supplementing courses with no-cost, openly licensed materials or to asking students to locate articles and evaluate other resources which are freely-available to them (at least while they are affiliated with the educational institution) is yet another way to test the waters. Many faculty have copious amounts of course notes already written. Turning these into a series of learning resources is a lot of work, but some faculty choose to do this. Other faculty involve students in creating openly licensed materials, such the student-created *Project Management for Instructional Designers* created in an Intro to Project Management course in 2011 and revised in 2012. This leads me into the next topic of open pedagogy.

Immersion into Open Pedagogy

As I mentioned earlier, my first instruction to Open Education at OpenVA 2013 included learning about Tidewater's Zero Textbook Degree and the reduction of student learning material costs. I knew that there was more to Open Education, this practice called 'open pedagogy' but it seemed inaccessible to me for a variety of reasons. However, that changed relatively quickly. In Fall 2015, I attended the Open Education Conference in Vancouver. It was there that 'textbook costs' and the ideas of 'open pedagogy' crashed together in a rather uncomfortable way. Criticism of the work of textbook affordability advocates seemed to go viral as seen through this collection of blogs. As a result, what I view as a synthetic divide was set up between proponents of open pedagogy and those advocating for open educational resources on the basis of reducing costs for students. To be clear, I think that both approaches are important and valid, and that one approach may be more attractive than the other depending on who you are. For students, cost is at times an enormous issue; many students are frustrated by requirement to purchase an access code to complete homework or to buy or rent a textbook they might not use in a course. While students are concerned about cost, faculty are concerned about fit, quality, and ways to facilitate meaningful student engagement in their courses. Both groups are important. Both approaches are valid. In fact, these two approaches are only two of many answers to the same question implied by Christina Hendricks, 'How can I make my course more open.'"

So what is open pedagogy? As Tom Woodward mentions in a _Campus Technology_ interview 'Open pedagogy is difficult … to crisply define.' Open pedagogical approaches seem to be characterized by increased student agency, relevance of course activities to the 'real world' (i.e., that they are public, useful, or valuable beyond getting a grade) or course activities and assignments which otherwise could not be implemented without open educational resources (items which adhere to the '5Rs' or are free to access, free to use, free to revise, free to remix, and free to redistribute). David Wiley describes open pedagogical practice as creating assignments that give value to the world which are not merely 'disposable assignments,' for example the _Project Management for Instructional Designers_ book mentioned earlier. Really, the sky is the limit when we think of the myriad of ways we can teach and learn using or creating public domain or openly licensed resources that we could not do with resources bound by cost or typical copyright restrictions.

I'm just starting to incorporate discussion of open pedagogical practices in workshops and instructional sessions. Faculty seem more to engage open pedagogy more quickly than they initially engaged with ideas of improving access by reducing cost. In the limited number of conversations I've had with faculty on this topic, there seems to be a lot of interest in planning ways to assign students more agency, making a course more public, creating assignments which are meaningful or useful beyond the course, or considering other ways to make courses more open or accessible. I'm not sure why this is. Perhaps it's an adjustment that seems more fun, interesting, and less overwhelming than selecting, adapting or authoring course content. Perhaps they are looking for an approach like this. I hope to more thoroughly explore this area in the coming months.

Conclusion

So, the panel discussion from the Spring of 2014 turned out fantastic. I was probably more relieved than anyone. Three faculty from two different institutions each discussed open and commercial works they created and why they created them. They also discussed what happens in a publishing ecosystem which combines both commercial/paid/royalty generating works and works which have no cost and are openly licensed. It was a fascinating discussion and a great first celebration of Open Education Week at Virginia Tech. This was a great start to the conversation about Open Education!

I'm hoping to keep the conversation going and to welcome additional faculty and students. I've spent quite a bit of the past year working to raise the level of awareness regarding open educational practices among administrators, departments, and groups that support teaching and learning. Several courses have committed to using open resources for the first time starting in Fall 2016. One of these is a pilot of the OpenStax *Biology* text in one of Virginia Tech's large enrollment courses. Another is a newly updated version of an openly licensed Business textbook.

Open Educational initiatives at the University Libraries will deepen as we plan to pilot a small grants program for open education resources in 2016 to 17. And work with students will be ever changing as some key student leaders and advocates graduate and others take their place. On the research front, two other librarians and I, via the Association of Research Libraries will publish a *SPEC Kit* on Affordable Course Content and OER in July 2016. The monograph is based on a survey of Association of Research Libraries (ARL) Library practices regarding Affordable Course Content and OER initiatives. I also hope to see some research results related to a free online but not-quite-fully openly licensed learning resource developed at Virginia Tech for a metabolic nutrition class. On a state and national level, I look forward to further service with OpenVA, the State Commission on Higher Education's Open Virginia Advisory Committee and service to the Open Textbook Network. It will also be exciting to see colleagues have the opportunity for further professional development and network building as part of the Open Textbook Network which the Virginia Virtual Library (VIVA) has joined. And the Open Education 2016 conference will be in Richmond, Virginia in November 2016. There are also a number of national and international projects and initiatives that coalesce with this work and are expanding access to research and scholarly works, such as Knowledge Unlatched, Open Library of the Humanities, Open Access Network, philpapers, OpenGLAM, and RightsStatements.org.

I have many hopes for faculty, administrator, and student engagement in making their courses less costly and more open. I hope that faculty leaders and administrators will be ready to voice public support for faculty engaging these practices. I'd very much like to see a campus-wide group form around open educational practices with leadership and broad engagement from teaching faculty.

In closing, I would be remiss not to mention that even though there is a great deal of work to be done that I'm very honored to be part of this movement. I'm grateful for mentors and friends who are so willing to share of their knowledge and expertise, faculty and administrators who engage these issues. Much of getting this work done depends on the continued sharing of both successes and failures, ideas and workplans across broad networks, and the continued engagement, persistence, and cooperation of faculty, students, administrators, librarians, instructional designers, information technology experts, and other concerned parties.

References

Amado, M., Ashton, K., Ashton, S., Bostwick, J., Nan, V., Nisse, T., Randall, D. (2016). *Project management for instructional designers.* Retrieved from https://web.archive.org/web/20160310104102/ http:/pm4id.org

Association of Research Libraries. (2016). *SPEC kits: Overview.* Retrieved from http://publications.arl.org/SPEC_Kits

Cabrera, N., Ostroff, J., & Schofield, B. (2015). *Understanding rights reversion.* Retrieved from https://web.archive.org/web/20160321233809/ http:/ authorsalliance.org/wp-content/uploads/Documents/Guides/Authors%20 Alliance%20-%20Understanding%20Rights%20Reversion.pdf

Exploring Innovative & Open Educational Resources. [Video file]. Retrieved from https://vtechworks.lib.vt.edu/handle/10919/46998

Grush, M. (2014). *Open pedagogy: connection, community, and transparency.* Retrieved from https://web.archive.org/web/20150915045516/ http:/ campustechnology.com/articles/2014/11/12/open-pedagogy-connection-community-and-transparency.aspx

Hendricks, C. (2014). *Presentation on open education at AAPT.* Retrieved from https://web.archive.org/web/20160414215952/ http:/blogs.ubc.ca/ chendricks/2014/08/08/open-ed-aapt/

Minding The Future. [Video file]. Retrieved from https://www.youtube.com/ watch?v=VJSkr1HA2mw&feature=youtu.be

Open Access Network. (2016). *Homepage.* Retrieved from: http://openaccess network.org/

OpenVA. (2016). *Homepage.* Retrieved from http://openva.org/

Open VA – Open Pedagogy / Curriculum Panel. [Video file]. Retrieved from https://www.youtube.com/watch?v=0yYAMrqk6NI&feature=youtu.be&t= 19m36s

Perry, M.J. (2012, December 24). The college textbook bubble and how the "open educational resources" movement is going up against the textbook cartel. [Web log comment]. Retrieved from https://web.archive.org/web/ 20140924011718/ http:/www.aei-ideas.org/2012/12/the-college-textbook-bubble-and-how-the-open-educational-resources-movement-is-going-up-against-the-textbook-cartel/

Perry, M. J. (2014, April 24). *Status 459395162032988160.* Retrieved from https://twitter.com/Mark_J_Perry/status/459395162032988160/photo/1

RightsStatements. org. (2016). Homepage. Retrieved from http://rightsstate ments.org/en/

Virginia.gov. (2016). *Open Virginia advisory committee.* Retrieved from http:// www.schev.edu/adminfaculty/advisoryCommittees.asp#OVAC

Walz. A. (2015, March 23). University libraries host open education week 2015. [Web log comment]. Retrieved from https://blogs.lt.vt.edu/ openvt/2015/03/23/university-libraries-host-open-education-week-2015/

Walz. A. (2016). *Virginia Tech Student Survey.* Unpublished manuscript, Department of Admissions, Virginia Polytechnic Institute and State University, Blacksburg, Virginia.

Web Archive. (2014a). *OpenVA live stream 2014.* Retrieved from https:// web.archive.org/web/20160414213654/http:/openva.org/openva-live-stream-2014/

Web Archive. (2014b). *Open education 2014.* Retrieved from https://web. archive.org/web/20141223164224/http://openeducation2014.sched.org/ event/db13570a116d6f2b64834835a9922711#.VxaiKGNcABg

Web Archive. (2014c). *Faculty survey finds awareness of open educational resources low.* Retrieved from https://web.archive.org/web/20141102002942/ http://www.babson.edu/News-Events/babson-news/Pages/141028-faculty-awareness-of-oea.aspx

Web Archive. (2015a). *Academics.* Retrieved from https://web.archive.org/ web/20160414212714/http:/web.tcc.edu/academics/zdegree/

Web Archive. (2015b). *Virginia tech: Cost of attendance.* Retrieved from https:// web.archive.org/web/20151023152536/http:/www.admiss.vt.edu/cost

Web Archive. (2016a). *Admissions and applications.* Retrieved from: https:// web.archive.org/web/20160324030737/http:/www.openuniversity.edu/ study/admissions-applications

Web Archive. (2016b). *About ds106.* Retrieved from https://web.archive.org/ web/20160330170415/http:/ds106.us/about/

Web Archive. (2016c). *A domain of one's own.* Retrieved from https://web. archive.org/web/20160326093739/http:/umw.domains/about/

Web Archive. (2016d). *Lauren Pressley.* Retrieved from https://web.archive.org/ web/20160321034104/http:/laurenpressley.com/lauren

Web Archive. (2016e). *OpenStax college.* Retrieved from https://web.archive. org/web/20160408155206/https:/openstaxcollege.org

Web Archive. (2016f). *Creative commons: About the licenses.* Retrieved from https://web.archive.org/web/20160413074147/https:/creativecommons. org/licenses

Web Archive. (2016g). *Creative commons: Attribution 4.0 international.* Retrieved from https://web.archive.org/web/20160413070720/http:/creativecommons. org/licenses/by/4.0/

Web Archive. (2016h). *Sparc library OER forum.* Retrieved from https://web. archive.org/web/20160414213909/http:/sparcopen.org/our-work/sparc-library-oer-forum/

Web Archive. (2016i). *OER consortium.* Retrieved from https://web.archive. org/web/20160408164804/http:/oerconsortium.org

Web Archive. (2016j). *Archived webinars.* Retrieved from https://web.archive. org/web/20160322203024/http:/oerconsortium.org/cccoer-webinars/

Web Archive. (2016k). *Twitter: Uberfacts: Status 611054921484472320*. Retrieved from https://web.archive.org/web/20160414213408/https:/twitter.com/uberfacts/status/611054921484472320

Web Archive. (2016l). *Searching for open materials*. Retrieved from https://web.archive.org/web/20160404114514/http:/libraryasleader.org/searching-for-open-materials

Web Archive. (2016m). Average estimated undergraduate budgets, 2015–16. Retrieved from https://web.archive.org/web/20160205031457/http:/trends.collegeboard.org/college-pricing/figures-tables/average-estimated-under graduate-budgets-2015–16

Web Archive. (2016n). *Higher education opportunity act*. Retrieved from https://web.archive.org/web/20160325222957/https:/www.gpo.gov/fdsys/pkg/PLAW-110publ315/html/PLAW-110publ315.htm

Web Archive. (2016o). *Academics*. Retrieved from: https://web.archive.org/web/20160420161814/http://httparchive.org/web.tcc.edu/academics/zdegree/

Web Archive. (2016p). *Open textbook network*. Retrieved from https://web.archive.org/web/20160330011436/http:/research.cehd.umn.edu/otn/

Web Archive. (2016q). *The virtual library of Virginia*. Retrieved from https://web.archive.org/web/20160306201611/http:/www.vivalib.org/

Web Archive. (2016r). *13ᵗʰ Annual open education conference*. Retrieved from https://web.archive.org/web/20160327035025/http:/openedconference.org/2016/

Web Archive. (2016s). *Knowledge unlatched*. Retrieved from https://web.archive.org/web/20160408191331/http:/www.knowledgeunlatched.org/

Web Archive. (2016t). *Open library of humanities*. Retrieved from https://web.archive.org/web/20160410073902/https:/www.openlibhums.org

Web Archive. (2016u). *PhilPapers*. Retrieved from https://web.archive.org/web/20160410155047/http:/philpapers.org/

Web Archive. (2016v). *OpenGLAM*. Retrieved from https://web.archive.org/web/20160311193059/http:/openglam.org

Wiley, D. (2016, April 7). Conversations prompted by #OpenEd15. [Web log comment]. Retrieved from https://web.archive.org/web/20160407060749/https:/storify.com/opencontent/opened15-post-conference-blogging

Wiley, D. (2014, March 5). The access compromise and the 5th R. [Web log comment]. Retrieved from https://web.archive.org/web/20160307064127/http:/opencontent.org/blog/archives/3221

Wiley, D. (2013, October 21). What is open pedagogy? [Web log comment]. Retrieved from https://web.archive.org/web/20160308004928/http:/open content.org/blog/archives/2975

Young, J.R. (2015, July 9). In students' minds, textbooks are increasingly optional purchases. [Web log comment]. Retrieved from https://web.archive.org/web/20150910151009/http:/chronicle.com/article/In-Students-Minds-Textbooks/231455/

How to Open an Academic Department

Farhad Dastur

Kwantlen Polytechnic University, farhad.dastur@kpu.ca

Editors' Commentary

Despite its wide-ranging benefits, the philosophy of Open represents change and, as a result, can easily threaten those who wish to maintain the status quo. In this chapter, author Farhad Dastur provides an insightful glimpse into a process of cultural change within an academic department—an organizational unit that he argues is effectively designed to resist change. In doing so, he provides a set of three practical recommendations for those interested in fostering change, including encouraging a departmental culture of openness, focusing on the quality of OER, and encouraging departmental control over OER.

'A ship in harbor is safe — but that is not what ships are built for.'
—John A. Shedd[1]

Introduction

This is the inside story of how my psychology department opened itself to the principles, practices, and possibilities of open education. We are two years into that story and far from finished. Nonetheless, I think this is a good time to pause, reflect, and share some insights that may help you as an agent of openness in your department.

Open education has the potential to transform the way we teach.[2] Unleashing that potential is imperative, but transforming complex institutions resistant to change is a wicked problem.[3] Universities have been around for almost a

How to cite this book chapter:
Dastur, F. 2017. How to Open an Academic Department. In: Jhangiani, R S and Biswas-Diener, R. (eds.) *Open: The Philosophy and Practices that are Revolutionizing Education and Science.* Pp. 163–178. London: Ubiquity Press. DOI: https://doi.org/10.5334/bbc.m. License: CC-BY 4.0

millennium and that persistence speaks to their remarkable immunity to new ideas and practices. Open education takes the spirit of sharing, creativity, and transparency and leverages those attributes with the flattening capabilities of the Internet, the portability of mobile computing, and the wider freedoms of flexible copyright and copyleft to achieve dramatic improvements in accessibility, content control, and creative collaboration. Against such lofty premises and promises, why has academia not embraced open education?

The answer requires an understanding of academic resistance to change. I propose just such an understanding: a theory born out of 15 years of observations as an educator, scholar, and academic administrator at Kwantlen Polytechnic University (KPU) in Vancouver, Canada. I conceptualize resistance as a structural phenomenon designed to protect the stability, integrity, and viability of academic departments and the faculty they serve. Complicating this conceptualization is the idea that organizations faced with transformative change are best understood as complex adaptive systems where the parts interact in unexpected ways. Linear prediction models are inappropriate in such systems; however, retroactive pattern sensing is possible.[4] With that in mind, I offer three pattern-based recommendations for opening your psychology department, namely, (1) encourage a departmental culture of openness; (2) focus on quality open educational resources (OER); and (3) encourage departmental control over OER.

Why Open?

Too many universities, and academic departments in particular, are closed in a world of blissful insularity, disciplinary elitism, strange and archaic traditions, reputational competitiveness, ivory tower detachment, ingroup vs. outgroup mistrust, resource competition, and epistemic fundamentalism.[5] This is the way it has been for much of the past thousand years. Indeed, so deeply are these features embedded into the fabric of our academic existence, that it is not uncommon to encounter colleagues who question whether this is even a problem. From the time we were undergraduates, through the long years of graduate training, and then as the professoriate class, we were all indoctrinated into the rigid rules and cultural codes of an institution older than the Crusades. Indeed, even the Collegiate Gothic architecture and religious iconography of many of North America's universities betray an Oxbridgian-inspired aesthetic preoccupation with gravitas, abiding permanence, and medieval heritage.[6]

Openness is grounded in the Enlightenment ideals of liberalism, freedom, citizenship, social progress, and transformation.[7] These ideals inform the open education, open access, open source, open science, open data, open design, and open government movements. For a summary of the contrasts between open and traditional education see the excellent chapter by Huitt & Monetti in this book. When I discovered open education—astonishingly late in my career—I recognized these ideals as the same ones I had always held as an educator, scientist, and citizen. I became a scientist because of a burning curiosity about the world; I became an educator because of an irrepressible desire to share that curiosity and the knowledge it led to. In that sense, my interest in openness was less like learning and more like remembering. A recollection of a time when the pursuit of knowledge was unmediated by formal education, unfettered by ideologies, and unencumbered with fears of failure, criticism, or dark sarcasm in the classroom. I still remember that evening in my childhood when I first saw Saturn and its rings through a telescope. All the fascinating astronomy I have learned since does not rival the purity and stillness of that perfect moment. That, too, is openness.

A Theory of Departmental Resistance to Change

My unit of analysis in understanding resistance to change is the academic department: the organizational unit primarily constituted by faculty, organized by discipline, and having significant control over disciplinary matters like curriculum. Departments represent the collective will of individual faculty in a given discipline (more on this fiction later). Much has been written on individual faculty resistance to change as well as institutional barriers to change.[8] Hopefully, my focus on the department offers different and useful insights.

The earliest medieval universities taught the Trivium and the Quadrivium; modern universities teach Nanotechnology, Postcolonial Literature, and myriad other courses. Though the courses have changed, universities' fundamental structures, governance models, and isolationist tendencies persist. For more than 95% of their history, universities were tasked with knowledge transfer from master to student. The hierarchical, parochial, and oligarchic governance and organizational structures that emerged over those centuries are still largely in play. This is either a fact of stunning consistency or appalling inflexibility. In his 1963 classic, *The Uses of the University*, Clark Kerr (2001: 115) observed that some 85 institutions in the Western world established by 1520 still existed in recognizable form and function including the Catholic Church, the British and Icelandic Parliaments, several Swiss cantons, and 70 universities.

> *'Kings that rule, feudal lords with vassals, and guilds with monopolies are all gone. These seventy universities, however, are still in the same locations with some of the same buildings, with professors and students doing much the same things, and with governance carried on in much the same ways.'*

Academic departments value tradition, reputation, autonomy, and disciplinary purity.[9] How many successful interdisciplinary programs exist at your institution? I recall a Faculty-wide curriculum committee meeting where a respected historian announced that only History's courses should be allowed to use the word 'history' in their titles. I've seen departments form alliances of solidarity when faced with the latest outrageous policy proposal from the Dean's Office, only to quietly form 'an understanding' with that same ambassador of 'The Dark Side' when seeking greater autonomy, resources, or preferential treatment.

Academic departments resist new ideas because they are designed to do so. It is no small irony that this happens in the very abodes where scholars tirelessly generate new ideas. However, new ideas when applied to departments signal a threat to the department's stability, interests, and self-preservation instincts. Like organisms striving to survive in a world full of change, challenge, and chaos, organizations are complex adaptive systems striving to survive their environmental vicissitudes. Several mechanisms exist for opposing change including procedural tactics at meetings; sending proposals for further study (read strangulation); writing letters of protest to the Dean; filing union grievances; finding common cause with similarly affected departments; and invoking a Fear, Uncertainty, and Doubt (FUD) campaign by suggesting that the change will threaten autonomy, job security, academic freedom, program quality, or institutional reputation.

An analogy drawn from immunology and informed by signal detection theory[10] may shed light into the nature of this change resistance. Consider how harmful antigens (e.g., bee venom) often trigger an antibody response that then neutralizes the antigen. I argue that new ideas, initiatives, and proposals are the antigens that trigger a departmental immune reaction. Think budget cuts or policy edicts from Administration that weaken departmental control. In an allergic reaction, the immune system is inappropriately triggered by harmless antigens (e.g., pollen). Symptoms like inflammation, sneezing, and tearing eyes follow. This reaction is called a false positive or Type I error: an inappropriate response to a non-existent threat. The opposite error, in which the immune system fails to respond to harmful substances, is a false negative or Type II error. Departments make false positive errors when they reflexively reject new ideas, initiatives, and policies. They make false negatives when they fail to respond to real threats. Figure 1 displays these four decision-making possibilities.

Which decision-making error is worse depends on what kind of organization you are and the nature of the threat. In organizations where stability and

	DEPT REJECTS IDEA	**DEPT FAILS TO REJECT IDEA**
BAD IDEA	Correct Decision	False Negative (Type II Error)
GOOD IDEA	**False Alarm (Type I Error)**	Correct Rejection

Fig. 1: Signal detection theory as applied to academic decision-making.

tradition are prized—universities for most of their +900-year-old history—we expect to see resistant decision-making that opposes most ideas (i.e., few false negatives and lots of false positives). The logic of this bias is that the consequences of failing to detect and respond to truly threatening ideas are far worse than the consequences of rejecting good ones.

Things began to fall apart in the 1960s. Universities had to contend with the unrest of the counterculture movement, the New Left, distorting marketplace influences, shifting societal expectations, and an increasingly technological and globalized world. Universities evolved into complicated and complex communities re-tasked with becoming transformative learning organizations that embrace change, serve society, and respond with agility to emerging opportunities. The governance models, power hierarchies, and pedagogical models that served them so well for the past nine centuries, now risked becoming the instruments of their own decline. The reflexive resistance to new ideas embedded into the DNA of departmental culture served well in maintaining a business-as-usual enterprise. But today business is unusual. Now, for perhaps the first time in history, a closed academic department is at risk of becoming a *closed* academic department.

Recommendation 1: Encourage a Departmental Culture of Openness

'The politics of the university are so intense because the stakes are so low.'

—Wallace Sayre[11]

In the spring of 2014, my colleague Rajiv Jhangiani, one of open education's most passionate advocates, challenged our department to embrace open education. In response, our departmental Teaching Excellence committee decided to discuss what all the fuss was about. My home was volunteered for the meeting. Over cups of French-pressed coffee and slices of freshly baked cake, we held our first open education *kaffeeklatsch* [kä-fē-, -kläch. noun. German, from *Kaffee* coffee + *Klatsch* gossip]. In retrospect, I consider this gathering the founding event in the project and process of opening our department.

In the two years since, we have made remarkable progress. Table 1 provides a timeline of developmental landmarks in the opening of our department. These landmarks span several themes including open education advocacy, teaching, research, presentations, course development, committee formation, policy review, and OER development.

I mentioned the kaffeeklatsch because the simple act of breaking bread during a meeting in a colleague's home is not so simple for some departments. As a naïve associate dean, I noticed departments divide into dueling dualities of sub-disciplines: Experimental vs. Clinical Psychology; Physical vs. Human

DATE	DEVELOPMENTAL LANDMARKS
April, 2014	Kaffeeklatsch to discuss OER adoption
July, 2014	Faculty from six universities in BC create 851 test bank questions during a 2 day Test Bank Sprint funded by BCcampus and NOBA
October, 2014	Symposium presentation at the Society for the Teaching of Psychology's Annual Conference on Teaching[12]
November, 2014	Formation of a departmental OER Committee
May, 2015	Faculty member gives the keynote at the Open Textbook Summit[13]
July, 2015	Presentation of faculty-led research at the 5th Vancouver International Conference on the Teaching of Psychology[14]
October, 2015	Faculty member (Levente Orban) begins work on a PsycWiki site including server installation and software configuration
October, 2015	Two faculty present at Open Access Week
November, 2015	Presentation of faculty-led research at the OpenEd 2015 Conference[15]
October, 2015	Two psychology faculty begin developing the WikiEducator-based open course, *Introduction to Psychology* for the Open Educational Resources Universitas (OERu)
December, 2015	Two psychology faculty join the institution-wide Open Studies Working Group
January, 2016	Launch of the department's Introductory Psychology OER Moodle website (Developers: David Froc, Richard Le Grand, & Kurt Penner)
January, 2016	OER Committee suggests ending the practice of using only one textbook for all sections of Intro Psychology and recommends that instructors be free to use either one traditional textbook or any open textbook
January, 2016	KPU becomes the first institution in BC to have over 100 course adoptions of open textbooks with a total cost savings to students of US$231,264
February, 2016	The Kwantlen Psychology Student Society attends a department meeting and urges faculty to consider adopting open textbooks when feasible

Table 1: Developmental Landmarks in the Opening of Psychology at Kwantlen Polytechnic University.

Geography; East Asian vs. South Asian Studies; and so on. Anthropology, once married to Sociology, now lived as a divorcée department with seven dependent members. Some departments refused to elect a chair. One historian refused to speak to another, while one criminologist ought to have refused to speak to another. Closed departments arise out of real or perceived mistreatment, toxic personalities, demoralizing budget cuts, ideological differences, incompetent leadership, competition for limited office and research space; and biased hiring decisions. And the situation at KPU is far from unique.

Academic tribalism replicates itself, fractal-like, up and down the institutional scale. Within psychology there are cognitive folks who look down on the personality folks. I've spoken to esteemed honeybee researchers that refused to speak to each other because of a bitter disagreement over which theory of colour vision best explained honeybee vision. And whom among us does not have a position on the qualitative/quantitative methodological divide or the tiresome nature/nurture debate?

Many formal and informal practices go into creating a transparent, collegial, collaborative, and healthy workplace. Here are some that characterize our department. On the whole, our department meetings are well run, often ending with faculty sharing a pint at a local pub. Faculty accomplishments are acknowledged in emails, at meetings, and in the departmental newsletter. An annual family-friendly retreat helps to build bonds of trust and future collaboration. Faculty provide input into the educational plan. Collegiality is a consideration in hiring decisions. One faculty member organizes the *Vancouver International Teaching of Psychology Conference* while another takes a lead organizing the *Connecting Minds Psychology Undergraduate Research Conference*. Faculty and students volunteer at both events. Mentorship is available for new faculty members to help them navigate the complexities of their budding academic careers. Pitched battles pitting faculty against students have played out on the badminton court, baseball field, and bowling alley. Intriguingly, recent research also suggests that the personality trait called openness to experience is positively correlated with faculty members' propensity to both create and adapt OER.[16]

Coming full circle, my first recommendation on how to open a psychology department is to first foster an open departmental culture. Create a culture where trust, communication, resiliency, and collegiality are the norm and watch your department show signs of spontaneous opening. And while that happens, please be sure to serve good coffee.

Recommendation 2: Focus on Quality Open Educational Resources

One of the biggest barriers to OER adoption, and open textbook adoption in particular, is the perception of inferior quality.[17] In one survey, 95% of Berkeley faculty identified 'Quality of content, including editorial review' as a necessary condition for open textbook adoption.[18] Reassuringly, the studies on faculty perceptions reveal that OER are typically viewed as equivalent in quality to traditional textbooks.[19]

When choosing a textbook, faculty decision-making is informed by several criteria including:

(a) The quality, readability, and organization of content.
(b) The ease of the adoption process.
(c) The quality and appeal of the illustrations and graphics.
(d) The types of ancillary supporting materials like test banks, PowerPoint slides, and Instructor's Manuals.
(e) The reputation of the author(s).

In terms of the latter, most psychology faculty would view favourably the traditional textbook, *Psychology*, authored by Harvard's Daniel Schacter, Daniel Gilbert, Daniel Wegner, and Matthew Nock (2014). How could an open textbook compete with such high caliber authors or, for that matter, with the power, prestige, and deep pockets of its dark overlord, Worth Publishers?

Ed and Carol Diener have an answer. The Diener's NOBA project[20] has developed an Introduction to Psychology OER with content modules written by widely respected scholars like Elizabeth Loftus,[21] Ed Diener,[22] Peter Salovey,[23] Roy Baumeister,[24] Henry L. Roediger III,[25] and David Buss.[26] Interested? And what if I told you that the online version of this OER was free and would save your students thousands of dollars? But wait, that's not all. If you adopt this OER right now, you will also receive a Creative Commons Attribution-Non-Commercial-ShareAlike license that permits the copying and redistribution of the content in any medium or format, and permits content adaptation, modification, and remixing for educational purposes.[27] Your only obligations are to give appropriate credit, provide a link to the license, indicate if you made any changes, and promise not to make a profit from the material.

Making faculty in my department aware that high quality OER already exist (e.g., NOBA, OpenStax, BCcampus); that they had been written and reviewed by recognized faculty; and that they are easy to access, helped overcome legitimate concerns about OER quality. Over the period 2013–2015, the percentage of our faculty teaching Introductory Psychology with an open textbook increased from 0% to 20%.

Recommendation 3: Encourage Departmental Control over Open Educational Resources

'A committee is a group of the unprepared, appointed by the unwilling to do the unnecessary.'

—Fred Allen[28]

Many activists use the master's tools to dismantle the master's house. However, in the effort to transform our department from a traditional model of education to a more open model, I have come to believe that the master's tools can also be used to renovate the house. One way to do that is to use and adapt pre-existing mechanisms that allow departments to function and project influence. These mechanisms include departmental meetings, reports, committees, educational plans and schedules, budgets, interactions with other institutional units, and the various formal and informal policies and practices that define the department's organizational culture.

One of the mechanisms of influence that we used was the creation of our own OER committee. This may seem counterproductive; after all, aren't faculty everywhere united in their disdain for committees and their procedural swamps of policy, protocol, and paperwork? Faculty chose their discipline out of love and interest, not a desire to sit on committees. They trained to be educators and scholars and then, one day, were bewildered to find themselves serving on 8, 9, or 10 committees. How could such a universally adopted mechanism for decision-making be so universally despised? Possibly because committees are remarkably effective at protecting and promoting individual and departmental interests.

Consider the following scenario in which a junior faculty member proposes a significant curricular change—let's call it 'Bright Idea'—to a departmental curriculum committee. Remember, this idea can trigger resistance in two ways: first, because it is a truly bad idea; second, because the committee is structurally biased to resist *any* idea. Committee members learn about the proposal and imagine all the *sturm und drang* it promises. In what ways could this coalition of the unwilling slow down or block the Bright Idea? After serving on hundreds of committee meetings, I now recognize at least seven subtle tactics which committees use to resist change. Let's call these tactics, 'The Seven Deadly Arrows of Committees' (Table 2).

First Arrow	Populate the committee with 'laggards' or 'late majority' adopters (Rogers, 2003)
Second Arrow	Demand unreasonable amounts of evidence
Third Arrow	Limit discussion and hold infrequent meetings on inconvenient days
Fourth Arrow	Relegate the Bright Idea to further study
Fifth Arrow	Declare that the Bright Idea is not part of the committee's mandate
Sixth Arrow	Oppose motions favouring the Bright Idea.
Seventh Arrow	Invoke a Fear, Uncertainty, and Doubt (FUD) campaign by suggesting that the Bright Idea will increase workload, threaten job security, dilute program quality, waste resources, or diminish academic rigour

Table 2: The Seven Deadly Arrows of Committees.

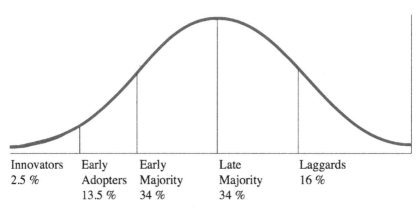

Innovators	Early	Early	Late	Laggards
2.5 %	Adopters	Majority	Majority	16 %
	13.5 %	34 %	34 %	

Fig. 2: Diffusion of innovation adoption curve.

This bleak destiny of bright ideas targeted for death-by-committee can be altered. To understand how, consider Rogers' (2003) diffusion of innovation theory which seeks to explain how innovations (ideas, behaviours, technologies) are adopted. When individuals in a social system, like a psychology department, are classified on the basis of their innovativeness, five normally distributed classifications emerge: innovators, early adopters, early majority, late majority, and laggards, with each category acting as an influencer for the next (Figure 2). The innovators (2.5%) are venturesome risk-takers while the early adopters (13.5%) are respected opinion leaders comfortable with change and uncertainty.

The curriculum committee members in our 'Bright Idea' scenario above are laggards and late majority adopters. The former are tradition-bound, conservative, and skeptical of change; the latter are change resisters who tend to adopt innovations only after successful adoption by the majority. Together, laggards and late majority adopters typically comprise 50% of a social group.

This theoretical model predicts that if a committee is strategically populated with innovators and early adopters, then there is a better chance for an innovation to be diffused through the entire social group. These two groups comprise 16% of the theoretical distribution, a number that falls very close to the 20% of our department members who currently serve on our OER committee and whom I consider change agents. Interestingly, a closer inspection of the open education developmental landmarks achieved by our psychology department (Table 1), reveals that every landmark achieved was by a member of our OER committee.

In the autumn of 2014, an interesting thing happened during a departmental meeting: I sought the department's blessings to use an open textbook for my Introduction to Psychology course. This request was without precedent as the department had an 'understanding' with a publisher to use only their Intro to Psychology textbook for at least three years. This exclusivity arrangement lowered the cost of the textbook and provided us with some scholarship money for students. Of course, using an open online textbook would cost our students nothing or, at most, the expense of a printed version (about 70% cheaper than the traditional textbook). The department's response was cautious and conditional: blessings would be given but evidence of the open textbook's relative efficacy should be collected. This request is reasonable but it simultaneously masks a characteristic of the pragmatic early majority adopters and the skeptical late majority adopters: interrogate new pedagogical practices and demand to be shown proof of their efficacy. In my experience, the reasonableness of this request is compromised by the observation that these same groups rarely subject their own pedagogical practices to equivalent scrutiny.

Notwithstanding this observation, the innovators and early adopters felt that data-driven arguments about the efficacy of open textbooks were important. And so several of the OER committee faculty developed a quasi-experimental research study to compare the efficacy of an Intro Psychology[29] open textbook vs. our traditional textbook.[30] The results of that study showed that, in terms of exam scores and qualitative student comments, the open textbook was at least as good as or better than the traditional textbook.[31]

Other made-in-psychology resources have further strengthened the open project including a Moodle-based OER of psychology learning objects and a PsycWiki. PsycWiki is a collaborative effort to create an open access textbook environment perpetually edited by students and faculty. This OER is modeled on wildly successful Chemistry LibreTexts library (2 million monthly visitors) developed by UC Davis' Delmar Larsen.[32] Important as these resources are, it is equally important that they are the product of faculty-driven initiative and collaboration. In this way, slowly but surely, the department comes to view these open education initiatives as its own and, ironically, begins defending them as part of its 'interests.'

Summary

Since their emergence in the Medieval Period, few institutions have resisted change more effectively or enduringly than universities. Academic departments with their closed systems, dysfunctional politics, disciplinary elitism, and internal mechanisms for opposing new ideas are a significant reason for this stasis. Academic departments resist change as a defensive strategy to ensure preservation. However, a signal detection theory analysis reveals that some of this resistance is simply the result of a bias to making Type I errors and, therefore, is unwarranted. However, by encouraging a more open departmental culture, by focusing on quality OER, and by encouraging departmental control of OER, this bias can be overcome and wider adoption can begin.

Our department's open project unfolded, and is unfolding, along multiple themes including advocacy, strategic committee formation, policy proposals, in-house research, course development, and OER co-creation and sharing. Two factors assisted Psychology's journey into openness. The first was a group of innovators and early adopters who found themselves in an already collegial departmental culture. The second was an institutional environment where key players were supportive of Psychology's initiatives including the Office of the President, the Vice President Academic, the Dean of Arts, the University Librarian (and librarians), other like-minded academic departments, and the psychology student society. Of course, change never happens in a vacuum, and it is also important to acknowledge that a network of stakeholders outside of our institution provided moral support, inspiration, financial aid, advice, and expertise. These groups include BCcampus,[33] the OERu,[34] the NOBA Project,[35] and OpenStax.[36]

Open education is a plea to the creators and illuminators of knowledge that it be shared; that it be open to co-creation, distribution, replication, modification, and integration; that collaboration and transparency be standard operating procedures; and that the barriers to knowledge access be dissolved. Inspired by this vision, several faculty members of KPU's psychology department began encouraging the opening of our department's culture, curriculum, and commitments. This was the story of the extraordinary opening of one ordinary psychology department. It is a story offered to you in the spirit of giving, perhaps the deepest act of openness imaginable.

Notes

[1] Shedd, 2006: 705.
[2] Atkins, Brown & Hammond, 2007; Jhangiani & Biswas-Diener, in press; Cape Town Open Education Declaration, 2007; National Knowledge Commission, 2007; OECD, 2007; Weller, 2013.
[3] Rittel & Webber, 1973.
[4] Kurtz & Snowden, 2003; Sargut & McGrath, 2011.
[5] Burton, 2004; Thorp & Goldstein, 2013.
[6] Meyer, 2013.
[7] Peters & Britez, 2008.
[8] Burke, 2011; Burke & Litwin, 1992; Coch & French, 1948; Dent & Goldberg, 1999; Lewin, 1951; Prochaska, Redding & Everts, 2009; Self & Schraeder, 2009.
[9] Burton, 2004; Kirschner, 2012; Rudolph, 1990.
[10] Peterson, Birdsall & Fox, 1954.
[11] Sayre, 2006: 670.
[12] Jhangiani & Dastur, 2014.
[13] Jhangiani, 2015.
[14] Le Grand et al., 2015
[15] Dastur et al., 2015
[16] Jhangiani et al., 2016; Peters & Britez, 2008.
[17] Arcos et al., 2015; Jhangiani et al., 2016
[18] Harley et al., 2010

[19] Bliss et al., 2013a; Bliss et al., 2013b; Feldstein et al., 2012; Hilton et al., 2013; Jhangiani et al., 2016.

[20] NOBA, n.d.

[21] Eyewitness Testimony and Memory Biases; Laney & Loftus, 2016.

[22] Happiness: The Science of Subjective Well-Being; Diener, 2016.

[23] Emotional Intelligence; Brackett, Delaney & Salovey, 2016.

[24] Self-Regulation and Consciousness; Baumeister, 2016.

[25] Memory (Encoding, Storage, Retrieval); McDermott & Roediger, 2016.

[26] Evolutionary Theories in Psychology; Buss, 2016.

[27] Creative Commons, n.d.

[28] Hopper & Davis, 1983.

[29] Intro Psychology, n.d.

[30] Myers, 2013.

[31] Dastur et al., 2015.

[32] Chemistry LibreTexts, n.d.

[33] BCcampus, n.d.

[34] OERu, n.d.

[35] NOBA, n.d.

[36] OpenStax, n.d.

References

Arcos, B. de los, Farrow, R., Pitt, R., Perryman, L., & Weller, M. (2015). *OER Research Hub Data 2013–2015: Educators*. OER Research Hub. Retrieved from https://oerresearchhub.files.wordpress.com/2015/09/educators_final_oerrhdata.pdf

Atkins, D. E., Brown, J. S., & Hammond, A. L. (2007). A review of the open educational resources (OER) movement: Achievements, challenges, and new opportunities. *The William and Flora Hewlett Foundation*, 1–84. Retrieved from http://www.hewlett.org/uploads/files/ReviewoftheOERMovement.pdf

Baumeister, R.F. (2016). Self-Regulation and consciousness. In R. Biswas-Diener & E. Diener (Eds), *Noba textbook series: Psychology*. Champaign, IL: DEF publishers. DOI: https://doi.org/nobaproject.com

BCcampus. (n.d.). Available at https://bccampus.ca

Bliss, T. J., Hilton, J., Wiley, D., & Thanos, K. (2013b). The cost and quality of online open textbooks: Perception of community college faculty and students. *First Monday, 18*(1).

Bliss, T. J., Robinson, J., Hilton, J., & Wiley, D. (2013a). An OER COUP: College teacher and student perceptions of Open Educational Resources. *Journal of Interactive Media in Education*, 1–25. http://doi.org/10.5334/2013–04

Brackett, M., Delaney, S., & Salovey, P. (2016). Emotional intelligence. In R. Biswas-Diener & E. Diener (Eds), *Noba textbook series: Psychology*. Champaign, IL: DEF publishers. DOI: https://doi.org/nobaproject.com.

Burke, W. W. (2011). A Perspective on the Field of Organization Development and Change The Zeigarnik Effect. *The Journal of Applied Behavioral Science, 47*(2), 143–167.

Burke, W. W., & Litwin, G. H. (1992). A causal model of organizational performance and change. *Journal of Management, 18*(3), 523–545.

Burton, R. C. (2004). Sustaining change in universities: Continuities in case studies and concepts. *Tertiary Education and Management, 9*(2), 99–116. http://dx.doi.org/10.7238/rusc.v2i1.250

Buss, D. M. (2016). Evolutionary theories in psychology. In R. Biswas-Diener & E. Diener (Eds), *Noba textbook series: Psychology*. Champaign, IL: DEF publishers. DOI: https://doi.org/nobaproject.com

Cape Town Open Education Declaration (2007). Cape Town open education declaration: Unlocking the promise of open educational resources. Retrieved from http://www.capetowndeclaration.org/read-the-declaration

Chemistry LibreTexts. (n.d.). Available at http://chem.libretexts.org

Coch, L., & French, J. R. P., Jr. (1948). Overcoming resistance to change. *Human Relations, 1*(4), 512–532.

Creative Commons. (n.d.). Available at https://creativecommons.org

Dastur, F., Le Grand, R., Jhangiani, R. S., & Penner, K. (2015, November). Introductory psychology textbooks: The roles of online vs. print and open vs. traditional textbooks. OpenEd 2015 Conference. Vancouver, BC.

Dent, E. B., & Goldberg, S. G. (1999). Challenging "resistance to change". *The Journal of Applied Behavioral Science, 35*(1), 25–41.

Diener, E. (2016). Happiness: The science of subjective well-being. In R. Biswas-Diener & E. Diener (Eds), *Noba textbook series: Psychology*. Champaign, IL: DEF publishers. DOI: https://doi.org/https://doi.org/nobaproject.com

Feldstein, A., Martin M., Hudson A., Warren K., Hilton J., & Wiley D. (2010). Open textbooks and increased student access and outcomes. *European Journal of Open, Distance and E-Learning, 2*, 1–9.

Harley, D., Lawrence, S., Acord, S. K., & Dixson, J. (2010). Affordable and open textbooks: An exploratory study of faculty attitudes. *California Journal of Politics and Policy, 2*(1). Retrieved from http://escholarship.org/uc/item/1t8244nb

Hilton, J., Gaudet, D., Clark, P., Robinson, J., & Wiley, D. (2013). The adoption of open educational resources by one community college math department. *The International Review of Research in Open and Distance Learning, 14*(4), 37–50.

Hopper, R., & Davis, L. J. (1983). *Between you and me: the professional's guide to interpersonal communication*. London, UK: Longman Higher Education. Retrieved from http://www.notable-quotes.com/a/allen_fred.html

Huitt, W. G., & Monetti, D. M. (in press). Openness and the transformation of education and schooling. In R. Biswas-Diener & R. S. Jhangiani (Eds.), *Open: The philosophy and practices that are revolutionizing education and science* (pp. xx–xx). London: Ubiquity Press.

International Journal of Wellbeing (IJW) (n.d.). Available at http://www.inter nationaljournalofwellbeing.org/)

Intro Psychology. (n.d.). Available at https://openstaxcollege.org/textbooks/ psychology

Jhangiani, R. (2015, May). *An openness to openness: The terrifying and liberating process of disrupting higher education.* Keynote presented at the 2015 Open Textbook Summit, Vancouver, BC.

Jhangiani, R. S., & Biswas-Diener, R. (Eds.). (in press). *Open: The philosophy and practices that are revolutionizing education and science.* London, UK: Ubiquity Press.

Jhangiani, R., & Dastur, F. (2014, October). *Opening up psychology: Adopting open textbooks, open pedagogy, and an open philosophy in the classroom.* Symposium conducted at the Society for the Teaching of Psychology's Annual Conference on Teaching, Atlanta, GA.

Jhangiani, R., Pitt, R., Hendricks, C., Key, J., & Lalonde, C. (2016). *Exploring faculty use of open educational resources at British Columbia post-second-ary institutions.* BCcampus Research Report. Victoria, BC: BCcampus. Retrieved from http://bccampus.ca/files/2016/01/BCFacultyUseOfOER_ final.pdf

Kerr, C. (2001). *The uses of the university.* (5th ed.). (p.115). Cambridge, MA: Harvard University Press.

Kirschner, A. (2012, April 13). Innovations in higher education? Hah! *Chroni-cle of Higher Education*, B6–B9. Retrieved at: http://chronicle.com/article/ Innovations-in-Higher/131424/

Kurtz, C., & Snowden, D. (2003). The new dynamics of strategy: Sense-making in a complex-complicated World. *IBM Systems Journal, 42*(3), 462–83.

Laney, C. & Loftus, E.F. (2016). Eyewitness testimony and memory biases. In R. Biswas-Diener & E. Diener (Eds.), *Noba textbook series: Psychology.* Champaign, IL: DEF publishers. DOI: https://doi.org/nobaproject.com

Le Grand, R., Dastur, F., Jhangiani, R., & Penner, K. (2015, July). *Using open textbooks for teaching introductory psychology.* 5th Vancouver International Conference on the Teaching of Psychology. Vancouver, BC.

Lewin, K. (1951). *Field theory in social sciences.* New York, NY: Harper & Row

McDermott, K.B. & Roediger III, H.L. (2016). Memory (encoding, storage, retrieval). In R. Biswas-Diener & E. Diener (Eds.), *Noba textbook series: Psychology.* Champaign, IL: DEF publishers. DOI: https://doi.org/noba project.com

Meyer, R. (2013, September 11). How gothic architecture took over the Ameri-can college campus. *The Atlantic.* Retrieved from http://www.theatlantic. com/education/archive/2013/09/how-gothic-architecture-took-over-the-american-college-campus/279287/

Myers, D. (2013). *Psychology* (10th ed.). New York, NY: Worth.

National Knowledge Commission, (2007). *Report of the working group on Open Access and Open Educational Resources.* New Delhi: National Knowledge

Commission, Government of India, 3. Retrieved from http://knowledge commission.gov.in/downloads/documents/wg_open_course.pdf

NOBA. (n.d.). Available at http://nobaproject.com

OECD. (2007). *Giving knowledge for free. The emergence of open educational resources.* Paris: OECD Publishing.

OERu. (n.d.). Available at http://oeru.org

OpenStax. (n.d.). Available at https://openstaxcollege.org

OpenStax College. (8 December 2014). *Psychology.* http://cnx.org/content/col11629/latest/

Peters, M. A., & Britez, R. G. (2008). Introduction. In M. A. Peters & R. G., Britez (Eds.), *Open education and education for openness* (pp. xvii–xxii). Rotterdam: Sense Publications.

Peterson, W. W., Birdsall, T. G., & Fox, W. C. (1954). The theory of signal detectability. *Proceedings of the IRE Professional Group on Information Theory, 4,* 171–212.

Prochaska, J. O., Redding, C. A., & Evers, K. E. (2009). The transtheoretical model and stages of change. In R. Glanz & K. Viswanath (Eds.), *Health behavior and health education: Theory, research and practice* (4th ed., pp. 97–117). San Francisco, CA: Josey Bass.

Rittel, H. W. J., & Webber, M. M. (1973). Dilemmas in a general theory of planning. *Policy Sciences, 4,* 155–169. DOI: https://doi.org/10.1007/bf01405730

Rogers, E. (2003). Diffusion of Innovations (5th ed.). New York, NY: Free Press.

Rudolph, F. (1990). *The American college and university.* Athens, GA: The University of Georgia Press.

Sargut, G., & McGrath, R. G. (2011). Learning to live with complexity. *Harvard Business Review, 89*(9), 68–76.

Sayre, W. (2006). Sayings. In F.R. Shapiro (Ed.), *The Yale book of quotations.* (p.670). New Haven, CT: Yale University Press.

Schacter, D. L., Gilbert D. T., Wegner D. M., & Nock, M. K. (2014). *Psychology* (3rd ed.). New York, NY: Worth.

Self, D. R., & Schraeder, M. (2009). Enhancing the success of organizational change: Matching readiness strategies with sources of resistance. *Leadership & Organizational Development Journal, 30,* 167–182.

Shedd, J.A. (2006). Sayings. In F.R. Shapiro (Ed.), *The Yale book of quotations* (p.705). New Haven, CT: Yale University Press.

Thorp, H., & Goldstein, B. (2013). *Engines of innovation: The entrepreneurial university in the Twenty-first century* [2nd ed.]. Chapel Hill, NC: The University of North Carolina Press.

Weller, M. (2013). The battle for open – a perspective. *Journal of Interactive Media in Education, 15.* DOI: https://doi.org/10.5334/2013–15

Case Studies

The International Journal of Wellbeing: An Open Access Success Story

Dan Weijers* and Aaron Jarden†

*University of Waikato, daniel.weijers@csus.edu

†Auckland University of Technology

Editors' Commentary

Academics have long had the advantage of access to university libraries and their expensive subscriptions to scholarly journals. Critics of traditional journal publishing have complained that placing science and scholarship behind a paywall limits its potential. One solution to this problem is the emergence of open access journals. In this chapter, authors Weijers and Jarden offer a case study of a platinum open access journal they founded: the International Journal of Wellbeing. In their discussion of this new journal they offer both philosophical and practical insights that guide their work. They also point to often overlooked issues regarding open scholarship. One of these is the huge numbers of unaffiliated faculty or faculty from non-Western universities, all of whom suffer barriers to access to expensive journals. The authors look to increasing openness of journals to solve this and other problems.

DW: There are not enough journals that publish interdisciplinary wellbeing research.

AJ: You're right. The few that do are choked up with submissions.

DW: We could create one, you know. There is free software for it.

AJ: That's a great idea. What would we call it?

How to cite this book chapter:

Weijers, D and Jarden, A. 2017. The International Journal of Wellbeing: An Open Access Success Story. In: Jhangiani, R S and Biswas-Diener, R. (eds.) *Open: The Philosophy and Practices that are Revolutionizing Education and Science.* Pp. 181–194. London: Ubiquity Press. DOI: https://doi.org/10.5334/bbc.n. License: CC-BY 4.0

DW: What about the *Australasian Journal of Wellbeing*?
AJ: Why not the *International Journal of Wellbeing*?
DW: But, could we really make it thoroughly international?
AJ: Sure, and it would be more fun. We should ask all of the people we really
         admire to be involved – and they're everywhere.
DW: But, why would they get on board with this?
AJ: I can see a lot of benefits, and I think they will too.
DW: Let's grab a coffee and figure this out…

The International Journal of Wellbeing: A caffeinated conception

In 2010, over coffee, this is roughly how we—the authors of this chapter—began exploring the idea of creating the *International Journal of Wellbeing* (IJW)[1]– an online only, interdisciplinary journal. We were motivated by our beliefs in the value of interdisciplinary wellbeing research and the importance of making useful academic research available to everyone. We also both love a challenge! At the time, Aaron held a junior faculty position at the Open Polytechnic of New Zealand, and Dan was a PhD student at Victoria University of Wellington. Less than a year later, on January 31st, 2011, the IJW launched its first issue. So, how did two young academics from the far corner of the academic world create an online only journal that, five years after launch and 15 issues later, is a respected and widely read journal (with unsolicited submissions from leaders in the field and nearly 500,000 full text article views)? Although hard work, quality relationships, and luck undoubtedly had roles to play, the key to the IJW's success was, and still is, its open access publishing model.

After our initial conversation, we conducted research into the various business models for journals. It did not take long to discover that the costs for publishing an online only journal were tiny. Overhead costs are about US$1,000 per year, and per article costs are about US$200, which includes professional copyediting, layout, and proofreading. We were astonished to discover this, and quite appalled at the current cost of journal subscriptions, article download fees, and one-off author fees for making individual articles open access in otherwise pay-for-access journals. For example, at US$39.95 plus tax per article (early 2016 price)[2], Springer would need five paying readers to recoup a reasonable per article cost of US$200. Even better for Springer, the authors could pay US$3,000 plus tax (early 2016 price) to make the article open access (free for everyone to read).[3] For an independent journal that uses professional copyediting, layout, and proofreading services, that US$3,000+ open access fee could cover the per article costs of fifteen articles. This begs the question, why so much? Springer and other academic publishers do add value, particularly with marketing, but they are also a profit-taking company with several layers of management. The truth behind most journal business models is that the vast majority of the hard work is done by academics. Thousands of hardworking academics find time to write, review, and edit for journals, often while juggling teaching and academic service responsibilities. Without those academics, the journals would fail, and with them, very little else is required for success. To be clear; it is not that Springer and other academic publishers do not add value—they do—it is just that they charge universities, academics, and others what we perceive to be a lot of money for added value that can easily be created in much cheaper ways.

The Internet has also played a huge part in all this, by making the dissemination of research orders of magnitude cheaper and more efficient than it was just 25 years ago; times have changed. One of the most important changes is the commitment of countless skilled people around the world, who create software and make it freely available for fun, and for the greater good. Governments and other organizations also play an important part when they fund open access and open source software initiatives. Most importantly for us, very high quality open source journal software is available for do-it-yourself journal publishers. We cannot thank enough the brilliant people at the Public Knowledge Project (PKP) and Open Journal Systems, as well as the US federal government for funding the opening up of academic research through the PKP. The software from Open Journal Systems allows even relatively techno-averse academics to do everything they need to in order to professionally manage and publish a quality academic journal.

The case for open access

With the knowledge that we only needed about US$5,000 per year to setup and run a successful online only journal, we realized that we did not have to approach journal publishers; we could publish the journal ourselves. The idea of open access publishing was appealing to us because it seemed fairer and more in line with our view of the point of academia – the production of useful information for all. Fully open access journals (that make all content free to everyone immediately upon publication) can be read by anyone with an internet connection and a computer. We believe that it is important that useful research publications can be accessed by as many people as possible, and as soon as possible. There exist many inequalities in the world, and unequal access to the most recent academic research, especially on wellbeing, is a particularly pernicious kind of inequality since it could exacerbate other forms of inequality. For this reason, we are proud that the research in the IJW has reached people in 185 nations around the world.

But who pays? Gold vs. platinum open access

Gold open access is not that open or fair

Making quality academic research freely available to everyone is a laudable aim, but even with academics volunteering to do most of the work, someone still has to pay for various parts of the publishing process. Many open access business models get the authors to pay. It does not seem fair to us to require authors to pay for the privilege of publishing in a journal, especially since they do most of the work in the research publishing process – writing research-based articles that extend the global body of useful information. This author pays kind of open access is sometimes referred to as 'gold open access,' but as we'll argue, it is gold in a credit card kind of way, not an Olympic medal kind of way.

Proponents of gold open access might argue that the authors never have to pay out of pocket because their institutions pay for them. But this reasoning is of little solace to unaffiliated scholars, academics at underfunded institutions, and academics outside of the sciences (where author fees are not the norm, and so institutional funds for publishing fees are less readily available). Some publishers waive the author fee for their fully open access journals to authors from low and lower-middle income nations. When publishing in Springer journals, for example, authors from Bangladesh do not have to pay a fee, but authors from India do. This is a good start, but many authors from India will not be able to get institutional funding, nor will humanities-based authors from many high-income nations. Therefore, even though gold open access articles and journals can be read by everyone, not everyone who might want to can publish in gold open access journals. In this way, gold open access is not completely open. This lack of access to gold open access journals can prevent many unfortunate academics from getting their research widely read, and widely cited, making it harder for them to progress in their careers. And, since this is through no fault of their own, we suggest that gold open access author fees can unfairly impact on some academics career prospects.

Gold open access might also be unfair to the funders of academic research. Passing the author fee on to academic institutions seems unfair to the institutions, since it is the academic institutions that pay the academics, enabling the writing of articles in the first place. While academic institutions are the main producers and consumers of academic research, the main funder of academic research is often governments. Governments subsidize academic institutions and fund research granting organizations, like the National Institutes of Health, because citizens generally value research-led education and the economic and other benefits of new academic research. As a general overview, governments fund the creation of research (via research grants and university subsidies) and the publishing of research (through open access fees or university library subscription fees). However, in some cases governments could pay four times for a published piece of research: First by subsidizing the university that pays the wages of the researchers to produce research. Second, through a directly funded research grant. Third, by paying the gold open access fee to make the research publically available. And fourth, by subsidizing the many university libraries that pay for the access to the pay-for-access journal that the author-paid-open-access article happens to be in. It is certainly possible that that the funders (mainly governments), the producers (mainly academics), and the consumers (mainly academics) of academic research are getting fleeced by publishers, which seem to add a small amount of value and take a huge cut of the profits.

Gold open access in action

Consider the example of the online only Springer journal Applied Physics B. In 2015, Applied Physics B published 66 articles, averaging nearly eight and a half pages in length. By our calculations, the per article cost for professional copyediting, layout, and proofreading could easily be as little as US$200 (our much longer articles cost us about US$200 each). So, the total per article cost for the whole year of articles could be as little as US$13,200. In 2015, 18 articles were gold open access, meaning that 18 authors (or their institutions or governments, depending on the particular circumstances) each paid up to US$3,000 plus tax to Springer. That's up to US$54,000 paid to Springer by authors in 2015 just for articles in Applied Physics B. Springer may well have higher per article costs than the US$200 we estimated, and they certainly have much higher overhead costs than us (e.g., upper management salaries), but it would be surprising if Applied Physics B didn't make Springer a decent amount of money *just* from the gold open access author fees and related institutional open access agreements. Of course, Springer also profits from selling annual subscriptions to Applied Physics B and individual articles from it. The current institutional annual subscription rate for the online only journal Applied Physics B is US$7,050 for the regular version and US$8,460 for the enhanced version.[4] We suppose that individual journal subscriptions are less common than subscriptions to bundles, so Springer probably receives less than US$7,050 per subscription to Applied Physics B. Even so, on the conservative estimate that 100 institutions and companies have a subscription to Applied Physics B, Springer receives hundreds of thousands of dollars in revenue each year through sales of subscriptions to the online only journal. As we said, Springer does add value with its publishing, hosting, and marketing process, but is it worth it? If one government or academic institution stumped up less than US$20,000 a year, then they could produce Applied Physics B to a similar standard. Why do governments and academic institutions (collectively) pay hundreds of thousands of dollars per year more than they need to? All told, gold open access doesn't seem all that open, or fair. It is OK, but something better is available, so golden open access is golden like a credit card, not like an Olympic medal.

Platinum open access is open, fair, and the way of the future

The editors of the IJW endorse platinum open access, which means we endorse publishing models that do not require readers or authors to pay. Platinum open access is more open than gold open access since, in platinum open access models, more people can contribute to the information. Platinum open access is also fairer than gold open access because academics who would otherwise have to pay their own author fees are not disadvantaged in this way, and because in platinum open access models, exorbitant publishing costs are not shouldered by those that have already funded the majority of the work. Platinum open access journals are usually funded by non-profit and charitable organizations, especially scholarly societies. Platinum open access journals can also be funded by academic institutions or self-funded through advertising revenue and donations. In general, academics are still doing most of the work, and they still pay most of the costs (or their institutions or governments do), but the costs are often dramatically lower. Some platinum open access journals are funded by academic societies and still published by for-profit publishers like Springer.[5] We think that this may not be the best choice, depending on how much commercial publishers charge for their services. A likely better choice would be to publish platinum open access journals independently or in association with a university library, established academic association, or similar institution. After the initial challenges of setting up, editors would have more freedom over format, and much lower costs. Platinum open access is more open and fair than gold open access, especially when it cuts out the expensive publisher-middleman. Platinum open access is to gold open access what a platinum credit card is to a gold one.

Just imagine what it would be like if governments decreed that all government funded research must be published in platinum open access outlets by the year 2026. Journals that are not currently platinum would investigate how to become platinum. For-profit publishing companies would have very little to offer them, unless they incorporate inordinate amounts of advertising in their publications. But not many companies want to advertise to academics since we are not a very lucrative target market. For-profit publishers might try to strike up direct pay-per article contracts with governments, but this price-setting might discourage quality as publishers pressure editors to be more generous with their acceptance rates. We think that a better solution would be for university libraries to cut their journal subscription budgets by 10% every year, and used that money to join with or create university presses and publish journals themselves. Given the declining interest in paper-based books and periodicals, librarians are in need of exciting new projects, so they should leap at the opportunity. There will likely be competition between top universities to secure the most prestigious journals, meaning that the editors of those journals should be able to secure ongoing top quality services and support for their periodical. The incentives will remain largely the same; authors will want to publish in the best journals, journal editors will want to attract the best research, and institutions will want to be associated with the best research, such as when they publish a top journal or employ someone who publishes in top journals. The main difference will be that all academic research will be completely open and the costs to the funders of research (mainly governments) should be lower because a profit-taking middleman has been removed from the process.

Why aren't governments already mandating that all government funded research must be published in platinum open access outlets? There are several possible reasons. Governments might be hesitant to be seen as restricting academic freedoms in any aspect of their work. Governments might not realize that most academics are unhappy with the current state of academic publishing. Governments might believe that forcing the academic market to comply with a platinum open access publishing mandate might create perverse incentives akin to the ones that have brought about the predatory open access journals that have proliferated in the last few years. Hopefully this book, and the advocacy of academics and other groups will help address these potential worries.

How open?: Creative Commons licenses

As we are sure readers are now well aware, not all open access is created equal. But the difference between gold and platinum open access is not the only important one. As academics as well as potential publishers, we realized that most academics are concerned to protect their intellectual property. For this reason, we wanted to make it clear that authors own the copyright to work published in the IJW, and we chose the Creative Commons license that gives authors the most protection, while ensuring the work can be used for all normal academic purposes without payment to the journal, the authors, or anyone else. As such the IJW uses the Creative Commons Attribution-Non Commercial-No Derivatives (CC BY-NC-ND) license. The attribution part of the license means that the work must always be explicitly described as being originally created by the authors. This is a standard protection for academic work. The no derivatives part of the license means that the work may not be re-versioned (imagine our open source software that you update and re-release) without the authors' permission. We believe that this offers an important protection for authors because it helps prevent third parties from changing the original work in ways that might reflect badly on the authors. Imagine changing a historical article about Hitler so that the authors appear to be endorsing, rather than merely reporting on Hitler's deeds. Our intention here was to respect the fact that many academics are hired and promoted (mainly) on the basis of the quality, quantity, and reception of the published work attributed to them. It makes sense for academics to care deeply about whether their work is attributed to them and whether it is being altered without their knowledge.

The non-commercial part of the license means that the work cannot be used for the purposes of making money (without the authors' express consent). This protects authors from third parties commercializing their ideas. The authors of this chapter have both been involved in commercializing academic research, and so we understand that many researchers wish to protect their valuable ideas, either to control how the research is put into practice, or to profit from it themselves. Some psychological scales, for example, are sold to mental health practitioners, generating profits for the authors of those scales. If a scale were published in an open access journal that did not have the non-commercial clause, then the scale could be used commercially by a third party without the permission of the authors. We believe that the non-commercial clause offers an, admittedly thin, layer of protection against third parties capitalizing off our authors work. We hope that this thin layer of protection might encourage authors to publish their research when it is finished, rather than after they have fully commercialized it. Importantly, though, we prefer this protection to be 'thin,' rather than a more robust copyright policy available from most academic publishers. Since this is a Creative Commons non-commercial license it does allow for research published in the IJW to be used by non-commercial groups without the authors' permission (as long as they attribute it to the author and do not make new versions of it). In line with our belief in the point of academic research being to create useful information for all (and essentially make the world a better place), we hope that the research we publish will be used (in a charitable or non-profit manner) to help people in need. Our main aim is to avoid the situation of some very useful research being published, but no one being able to put it to use because the authors have not put it to use.

So far, no authors have complained about our use of the Creative Commons BY-NC-ND license. A few authors have been pleased about how easy it is for them to use their work in other ways. For example, authors can host the original or adapted versions on their own website, reproduce the article in an anthology or monograph, or even print and bind it nicely and sell it as 'a good birthday present for the intellectual in your life.' All we ask is that they acknowledge that the IJW was the original publisher.

The prestige barrier and the open access solution

The prestige barrier

Given all of the arguments above, we knew that we wanted to create an inter-disciplinary online only platinum open access journal on wellbeing (broadly construed, including disciplines such as philosophy, psychology, economics, and sociology). We also desired for the journal to be an exceptional one; a high quality well-respected journal that leading academics would be proud to publish their work in. As junior academics, we were confronted with a huge prestige problem; why would the best scholars in the field want to join our editorial team or submit their research to us? These concerns cannot be under-stated. While all academics feel the pressure to publish in the most well-known journals, young academics feel it acutely. In many disciplines and universities, research publications count for hiring, tenure, and promotion only if they appear in an often implicit, but usually set, list of journals. New journals cannot be on those lists straight away, and are only likely to make it onto those lists after they prove themselves through publishing top quality research. Soliciting excellent research is difficult when your journal is not already on that list. Academic prestige, the promise of academic excellence achieved by association with perceived academic excellence, is the key to attracting those initial top quality submissions.

Unfortunately for us, academic prestige also tends to favor established ways of thinking and operating. All of our buzzwords (open access, interdisciplinary, online only, and wellbeing) are relative newcomers to academia, which made us think that they were more likely to raise 'prestige red flags' than our chances of getting the IJW off the ground. Both being educated and working in New Zealand, rather than at an Ivy League university in the US, we did not have institutional prestige to leverage. Aaron had a few contacts with excellent psychologists working on wellbeing, but other than that, there was no reason to think we had the resources to resolve our prestige problem. If we were flexible on our platform open access status, then we might have been adopted by an established journal publisher. Being associated with Elsevier or Taylor and Francis might have allayed fears that the IJW would only be read by people who accidentally found the page via a procession of typos. However, our principles insisted that we did not give up on platinum open access. So, armed with just a few contacts, and an overabundance of naïve optimism and caffeine, we pushed ahead with our idea.

The open access solution

Before too long, we were joined by our third co-editor, rising star in economics, Nattavudh Powdthavee. We then created a 70-page business plan that stressed the low cost of online publishing (even of the highest quality), and the IJW's main point of difference – being fully open access. Then began the nerve-wracking process of contacting our academic idols, explaining the rationale and the mandate of the IJW, and inviting them to join our editorial team. To our surprise, nearly all of the academics we contacted enthusiastically agreed to not just put their names to the IJW, but also to offer their time and effort in many different ways. In a matter of weeks, our editorial and advisory boards were brimming with many of the best established and up-and-coming researchers in the field. All of a sudden, we found ourselves with the support of people who could resolve our prestige problem, and provide us with invaluable advice on editing and publishing.

But why were they all so eager to get on board with us, two novice academics from New Zealand? For the vast majority of our now colleagues, and especially the more established ones, the main attraction was our fully open access publishing model. Many of them shared our disgruntlement at some of the academic publishers' prices and other practices. They saw supporting the IJW as something they could do to help turn the tide back toward the ideal of making useful information available to everyone, and not double or triple-charging the academic institutions aiming to achieve this goal. World renowned wellbeing expert and IJW advisory board member, John Helliwell, for example, made it very clear to us that his tremendous efforts to help establish and promote the IJW were motivated by our commitment to platinum open access publishing.

With the support of John Helliwell, and other leaders in the field of wellbeing research, we were able to attract enough funding to cover the IJW's start up and operating costs for at least eight years. Initially, technical and webhosting support was provided by the Open Polytechnic of New Zealand, and funding for professional copyediting, layout editing and proofreading was generously provided by the Vic Davis Memorial Trust (a mainly community-based mental health funding organization in New Zealand). After two years of invaluable service, Nattavudh relinquished his co-editorial role, and was replaced by Stephen Wu from Hamilton College in New York. The operations of the IJW are now generously funded by Hamilton College, who showed interest in the journal because of its topic, early success, and open access status. Shortly after, Lindsay Oades, psychologist at the University of Melbourne Australia, also joined our co-editorial team and brought with him much policy and process experience. After we attracted so many leading wellbeing scholars to help run the journal, it was clear that the prestige problem had been overcome, and perhaps even turned right around. In fact, reflecting back to the very first issue of the IJW, we attracted such esteemed scholars as Martin Seligman, Fred Bryant, John Helliwell, Erik Angner, and featured an interview with Nobel laureate Daniel Kahneman. As such, institutions have become enthusiastic about aligning with and funding the IJW, especially considering it is cheap to do so. As a result, we have published 113 articles and reviews over the last five years.

While many things go into creating a successful journal, the moral of this story is that a commitment to platinum open access publishing is what made the difference between us creating a one-issue-wonder and the globally read and increasingly influential IJW.

Has the IJW made a difference?: Opening up wellbeing studies

Affiliation troubles

Although some measures of the IJW's impact have already been mentioned, we'd like to emphasize the many ways in which the IJW has helped to open up wellbeing studies. In the publish-or-perish world that emerging academics struggle to survive in, losing an affiliation with an academic institution is often the death knell for an academic career. Unaffiliated scholars are less likely to be invited to conferences, have their manuscripts accepted for peer review, and be invited to interview for academic positions. To make matters worse, they also have their access to the latest research nearly completely cut off. They will unlikely be able to afford to read many articles, especially at approximately US$40 an article. Sure, the unfortunate unaffiliated academics could approach authors directly, and ask for pre-prints, but if they are conducting serious research then they would be sending such requests almost daily. It might be argued that this inconvenience only affects a small number of people, since the half-life of unaffiliated scholars is relatively short. However, many affiliated scholars find themselves in a similar situation. Academics affiliated to institutions outside of the West, and even the less established ones in the West, will rely on libraries that are struggling to keep up with the rising cost of bundles of journals. A recent Harvard University memo revealed that their library is struggling to pay for subscriptions to scientific journals, which now cost upwards of US$3.5million.[6] Furthermore, a very similar story can be told about unaffiliated and less-fortunately-affiliated scholars and author fees.

The beauty of platinum open access publishing is that these problems disappear. As long as the article is of sufficient academic merit, it can be published in the IJW (18% of submitted articles were accepted in 2015), and then disseminated around the world for everyone to read without requiring the author to pay a US$3,000 gold open access fee. For example, emerging academic Rachel Dodge, and her co-authors, submitted an article to us from a partially affiliated position. At the time Dodge submitted 'The Challenge of Defining Wellbeing' to the IJW, she was a part time PhD student at Cardiff Metropolitan University in the United Kingdom. Since we did not evaluate the submission on its authors' affiliations, or its lead author's position in the academic hierarchy, and since there were no submission or author costs, Dodge encountered no barriers to publishing her work with the IJW. After review and subsequent revisions, the paper was published in 2012. As a very junior academic—a part time student— Dodge did not expect that her research would make much of an impact in her own country, let alone the world. However in a little over three years, her article 'The Challenge of Defining Wellbeing' has been viewed over 50,000 times in at least 96 countries, including multiple views in places such as Iran, Rwanda, and Peru. The combination of the IJW's online and free and unrestricted access policies truly makes the academic research we publish available around the world. Furthermore, most academics do not just want their research to be read by a lot of people, they also want their research to impact the relevant scholarly debates. Fortunately, given academics' propensity to do their research online using search engines like Google Scholar, open access research is readily accessible to all academics, including those at privileged research institutions. The truth of this is perhaps best evidenced by the fact that (according to Google Scholar) 'The Challenge of Defining Wellbeing' has been cited 151 times in just over three years.[7]

Opening up the discipline to new readers

To further take advantage of the IJW's broad accessibility, our second issue was aimed at scholars from all disciplines and educated non-academics. 'Felicitators,' as it was called, was designed to act as an entry point to academic research on wellbeing for academics and laypeople alike. The Felicitators issue effectively made interdisciplinary research on wellbeing more open by placing an even greater emphasis on doing away with unnecessary disciplinary jargon and focusing on real world examples. The issue included articles from a diverse range of authors, including an artist, a philosopher, a monk, a historian, and social scientists. The issue also covered an eclectic range of topics, including Montessori education, Dr Seuss's The Lorax, a Singaporean prison, a music teacher's inclusive approach, Bruder Klaus on peace and war, and an investigation of whether Moses was happy. We were very pleased to enable the publication of academic work that we could whole-heartedly direct our non-academic friends and family towards, in full knowledge that the articles would be accessible to them (in both senses of the word; i.e., free and understandable). We liked the Felicitators idea because it was open in these ways, but we could not have published it if the IJW was not open as well. As independent publishers, we the co-editors of the IJW, were free to decide what we would publish, and in what format we would publish it. We are free to open up existing publishing practices and help produce issues that might not otherwise be published in a venue that is accessible to academics and lay people.

Open access: Into the future

Could the IJW be more open?

When setting up the IJW, we also considered other ways that the journal might be more open. We currently encourage authors to provide data and qualitative transcripts, which we publish as supplementary files. When authors publish their source data and qualitative transcripts, other researchers can use and challenge them, moving the boundaries of shared knowledge more quickly and transparently. We decided not to mandate the publication of source data in the end because we feared that the majority of authors are not ready to share their data and open themselves up to criticism and the possibility of being scooped on future publications. However, we hope that the funders of research will start to mandate that all data sources are published in open access venues alongside the research articles based on them.

The IJW editors also considered an open review policy, according to which reviews of articles are published along with the articles themselves. Sometimes reviewers names are included, and sometimes not. The transparency benefits of these kinds of policies are very appealing. The robustness of peer review would be visible to everyone. It is also argued that open review could help reward the currently thankless task of reviewing the manuscripts of others. Again, we decided that the key stakeholders were not ready; we believed that most academics are so busy that they would be much more reluctant to agree to review an article publically. Being so busy means that most academics will worry that the review will take longer because producing a mistake-free review that is polished enough for public consumption will take twice as long as a regular review would. Furthermore, if reviews have the reviewers' names attached, this could reflect very badly on them if they make a mistake, again putting pressure on them to spend much more time on the review. All this extra time that reviewers might spend on reviews seems like a point in favor of open reviews, and it is from the big picture perspective, but it is seen as a disincentive for most reviewers. We figured that academics are already less likely to review for a new journal, so we decided against applying this further pressure on reviewers. It seems that only a massive re-organizing of several academic institutions could help resolve this problem, something like centralizing all potential reviewers and balancing their reviewing workloads, but such huge changes might introduce new problems.

The IJW could be more open, and being more open in the ways just mentioned would probably provide the most benefit. Unfortunately, our view is that at the present time academics themselves do not quite seem ready for this level of openness. Perhaps in the next decade we will see a cultural shift regarding this extreme openness, but it is more likely that academics will need to be nudged or coerced by funding institutions before they become more receptive to opening up peer review and their source data. When we feel like our most important stakeholders are ready, the IJW will happily adopt these more open policies.

What now for the IJW?

The IJW now has a great team of enthusiastic people involved, the software is robust and well managed and maintained, funding has been obtained for at least the next three years, and the disciplinary reach of the journal and the impacts of the work it publishes are further increasing. Looking into the future, we can see that the IJW, buoyed by its commitment to open access, will go from strength to strength. It is no wonder that we have been approached by major journal publishers who would like the IJW on their books. But, the IJW is not a commercial venture; it is an academic venture with the aim to disseminate useful information on wellbeing as widely and as openly as possible, and a commitment to open access.

Naturally, we still have many challenges ahead, such as keeping a pace with publishing trends and newer tracking technologies, sheer volume of submissions, more thoroughly indexing the journal with databases and search engines, and making sure we are not mistakenly thrown in with the new explosion of fake open access journals. But as for now, the IJW is still running well on all the hard work and enthusiasm of the open-access-inspired academics and others involved, and the coffee of course, which continues to flow. We would like to take this opportunity to thank all of the academics who find the time to help produce and disseminate quality research through platinum open access channels – together we are making a positive difference.

Notes

1. International Journal of Wellbeing, n.d.
2. Prices are readily available on www.springer.com. Visit any normal ('Open Choice') Springer journal when you do not have a subscription, and check the price to download an article.
3. Open Choice prices and information are readily available here: http://www.springer.com/gp/open-access/springer-open-choice.
4. See all Springer journal subscription prices here: https://www.springer.com/gp/librarians/journal-price-list.
5. See http://www.springer.com/gp/open-access/springer-open.
6. See, for example, https://www.theguardian.com/science/2012/apr/24/harvard-university-journal-publishers-prices.
7. The Web of Science citation tracker, which includes only select scientific journals, notes 28 citations of 'The Challenge of Defining Wellbeing' as at 12 March 2016. The 151 citations in Google Scholar was also as at 12 March 2016.

Iterating Toward Openness: Lessons Learned on a Personal Journey

David Wiley

Lumen Learning, david.wiley@gmail.com

Editors' Commentary

For nearly 20 years David Wiley has been on the frontlines of the open educa-tion movement, working on tools, licenses, infrastructure, research, and advocacy. In this chapter, David shares personal and hard-won insights from his mission to implement the ideas and promises of Open, including the power of combin-ing digital content with open licenses, the pointlessness of producing OER that is never reused, the kind of change required to realize the potential of Open, the need to redefine OER quality in terms of its effectiveness, and the importance of addressing specific problems. The chapter concludes with a commentary on the infrastructure that is necessary to truly expand educational opportunities and potential.

I've spent my entire career watching very smart and well-meaning people claim that the unique features of *their* repository of open educational resources (OER) (or learning objects the decade before that) will finally result in sig-nificant teacher use, or that *their* authoring tools are so wonderfully easy to use that they will create a breakthrough in faculty use of OER. While some repositories and some authoring tools are certainly easier to use than others, a website simply cannot overcome the gargantuan inertia and imprinted behav-iors associated with textbook use in higher education. And that's my current

---

How to cite this book chapter:
Wiley, D. 2017. Iterating Toward Openness: Lessons Learned on a Personal Journey. In: Jhangiani, R S and Biswas-Diener, R. (eds.) *Open: The Philosophy and Practices that are Revolutionizing Education and Science.* Pp. 195–207. London: Ubiquity Press. DOI: https://doi.org/10.5334/bbc.o. License: CC-BY 4.0

goal – to replace the expensive, rights-restricted textbooks currently in use in higher education with OER. OER are materials that meet the criteria of free plus permissions – they are (1) freely available and (2) come with an irrevocable grant of permission to engage in the 5R activities – retain, reuse, revise, remix, and redistribute. It's true that better tools will help move the field forward, but tools can't get us all the way there (or even close to there, honestly). To borrow an analogy from Hubert Dreyfus, trying to influence the textbook adoption process of the nation's faculty by making a website is a bit like trying to get to the moon by climbing trees. Yes, you can make some initial progress that moves you a slightly closer to the goal and feels encouraging, but the whole approach is doomed from the beginning.

In this chapter I share some of the lessons I've learned as I've slowly iterated toward openness over the last twenty years, in hopes that they can help move higher education toward OER more quickly and efficiently.

Finding Open

In 1997 I finished my BFA in Music at Marshall University and was working as the institution's first webmaster. No one knew what a webmaster was supposed to do, myself included, and I enjoyed broad latitude in my day to day activities. In addition to chairing the committee that determined who owned the rights to content created by faculty for online courses (who better than the webmaster, right?), I spent a lot of time exploring new technologies.

One day I was playing around with Javascript trying to build a calculator that could be embedded in a webpage. I will remember that afternoon for the rest of my life. It suddenly occurred to me that once this calculator was created and published online, the whole world could use it. Well, more than that – the whole world could use it *at the same time*. That affordance of being digital made it critically different from a physical calculator that only one person at a time can use. Of course economists and others had understood the difference between rivalrous and nonrivalrous resources for years, but this discovery was completely new to me.

It was quite vogue at the time among Slashdot-reading free software advocates to think poorly of Bill Gates. I connected my calculator discovery to his incredible wealth immediately. Silently shaking my head, I thought something along the lines of 'This is how you become a billionaire – create something that costs nothing to copy and then sell copies for US$150 each. It's like printing money!' Then the better angels of my nature took the mic and suggested, 'The other side of the coin is this – once you create something digitally, it can be used by everyone around the world at no additional cost.' Not just tools like calculators, I realized, but syllabi, articles, chapters, entire journals and whole textbooks… This realization hit me like a bolt from the blue. It was almost like in a movie, where the clouds part and a ray of sunshine breaks through. In that moment I had the undeniable impression that because I understood it, I was responsible to act on this knowledge.

In 1998 I left Marshall and headed to Brigham Young University to take a PhD in Instructional Psychology and Technology. Earlier that year, Eric Raymond, Bruce Perens, and others had proposed the term 'open source' as an alternative to 'free software.' I was particularly struck by the pragmatism of their arguments. Cutting the grass of the tiny lawn next to our apartment in Orem that summer, I realized that open licenses were the key ingredient I was missing. *Digital formats make the broad sharing of educational materials technologically possible; open licenses make the broad sharing of educational materials legal.*

Making Open

And so I set out on a journey with a general direction but without a specific destination. The power and pragmatism of the open source model, the almost magical nature of digital content, and the way these combined to catalyze the wonders of the internet fascinated me. It seemed like an incredibly promising approach that could transform education and provide huge benefits to people in the process.

I spent the first ten years of this journey working with a number of collaborators to build very specific bridges across very specific chasms between where we were and where we wanted to be in terms of bringing the power of open to education. This included creating and propagating the 'open content' meme, creating the first open licenses for something other than software, including the Open Content License and Open Publication License (later superseded by the Creative Commons licenses), persuading individuals and institutions to begin using these licenses to share their open educational resources, and providing technology infrastructure to university-based OpenCourseWare initiatives around the globe. Years later, thousands of universities, individuals, and organizations were sharing openly licensed educational materials online.

To my dismay, however, almost no one was using OER in formal settings. Yes, projects like MIT open courseware (OCW) published evaluation reports showing that individual learners from the around the world were coming to their website and learning things. Many of their stories were incredibly moving. But the open education movement, as we were calling it, wasn't actually impacting formal education. People were openly licensing materials left and right, but faculty continued to adopt expensive commercial textbooks for their courses. For a year or so it felt like every week saw another major OCW announcement from a major university around the world. But while they all wanted to *publish* OER, no one wanted to *use* OER. In retrospect, perhaps this was because the early participants were all extremely prestigious schools. The rivalries between these schools being well understood, would we really expect faculty from Ivy League A to reuse anything created by faculty from Ivy League B?

This situation frustrated me to no end. What was the point of openly licensing educational materials if no one was going to use them? (Early on, the answer to this question was 'publicity.') And why were people calling this pattern of behavior 'sharing?' If I offer you some of my French fries, but you don't take any, have I shared with you? No. Rather than calling it open sharing, the first decade of work in the open education movement would be more accurately characterized as open offering. *If no one was going to reuse these openly licensed materials, the whole exercise was literally pointless.*

Reusing Open

In 2007 a graduate student walked into my office at Utah State University to inform me that Utah state law had changed recently, making it legal to open fully online charter schools. I responded with a confused, 'That's interesting... thanks for sharing.' He pushed ahead, 'I think this is a great opportunity for you to put into practice all your fancy theories about open content and education.' He didn't quite say 'put up or shut up,' but close enough. I began discussing the idea in earnest with colleagues in the Center for Open and Sustainable Learning (COSL), the research unit I founded and directed at USU. This seemed like a great opportunity to make progress toward our stated mission:

> At the Center for Open and Sustainable Learning, we believe that all humans beings are endowed with a capacity to learn, improve, and progress. Educational opportunity is the mechanism by which we fulfill that capacity. Therefore, free and open access to educational opportunity is a basic human right. When educational materials can be electronically copied and transferred around the world at almost no cost, we have a greater ethical obligation than ever before to increase the reach of opportunity. When people can connect with others nearby or in distant lands at almost no cost to ask questions, give answers, and exchange ideas, the moral imperative to meaningfully enable these opportunities weighs profoundly. We cannot in good conscience allow this poverty of educational opportunity to continue when educational provisions are so plentiful, and when their duplication and distribution costs so little. (COSL website)

We decided to create something new – a major OER initiative that would not produce any OER, but instead would be dedicated to reusing OER produced by others.

This was the genesis of the Open High School of Utah (OHSU; which later changed its name to Mountain Heights Academy), a charter school whose charter documents commit it to using OER for its core curriculum materials instead of commercial resources. This turned out to be much harder to do than it sounded. Yes, the global collection of OER was still relatively young in the mid-2000s, but more importantly the school's new faculty, administration, and board would need *significant*, ongoing professional development to understand what it was we were trying to do and why it was important.

I learned first hand that this degree of systemic change – changing the fundamental way schools and teachers find, procure, use, and continuously improve curriculum resources – is not the kind of change you can create by conducting a workshop. It's also not the kind of change you can create by simply developing a tool with a simpler user interface (UI) or a bigger library of content. *This degree of change requires sustained attention by people who care deeply about it succeeding, and nothing short of that will work.*

Comparing Open

As I talked with people about our early success with OHSU, I began to hear the discourse around OER shift. A few years earlier, people demurred when you told them about OER because they were certain none existed in their discipline. As we were able to demonstrate that sufficient OER did exist, the excuse changed to concerns about quality. The old notion that 'you get what you pay for' was simply too deeply engrained in people. Yes, I could convince someone that OER existed for their courses, but they couldn't believe that anything freely available could be worth their time.

In 2008 I decided to shift strategies and institutions. I moved from Utah State University to Brigham Young University, and decided to intensely focus my research on cutting the legs out from under the quality arguments around OER. I hoped to be able to successfully fight intuition with data.

The first problem with fighting the public's perceptions of 'quality' of educational materials is that the public is *completely and utterly wrongheaded* in their thinking about quality. Let me explain.

As success at OHSU translated into adoption of OER in place of commercial science textbooks in traditional Utah high schools and middle schools, my graduate students and I engaged in several research studies. At the same time evidence was emerging that students who were assigned OER did as well as their commercial textbook using peers on the state's standardized science tests, a Brigham Young University (BYU) graduate student completed a master's thesis examining the quality of the OER they were using. She concluded that the OER were of lower quality compared to commercial materials, based on a structural and aesthetic review of the OER. There were problems with the layout and graphic design of the OER, there were copyediting shortcomings, there were pixilated images in the text, etc. By any aesthetic or 'production values' measure, the OER were lower quality than the materials being provided by publishers. However, students were learning the same amount – in some cases more – from the OER.

This contradiction inevitably lead me to ask, 'what is quality?' When given the choice between materials that are beautiful but result in lower learning and materials that are far less beautiful but result in better learning, which will we call 'higher quality?' (If you had a beautiful hammer that drove nails poorly, and an ugly hammer that drove nails effectively, which would you call higher quality?) What is the primary purpose of educational materials? Is it to win a beauty contest, or to support learning?

Traditional publishers were quick to latch onto and propagate the 'OER are low quality' message. For traditional publishers, quality was purely a function of editorial process and had literally nothing to do with student learning. By making 'quality' equivalent with expensive graphic design, editorial, photography, other artwork, and other creative processes publishers sought to set 'quality' out of the economic reach of fledgling OER initiatives. By shouting from the rooftops that the quality of an educational resource ought to be judged by the learning it facilitates, I hoped to change the fight into one that OER could win. This is why, when talking about educational materials, I typically refuse to use the word quality and instead use *effectiveness*. If materials are less effective, who cares how beautiful they are? If they are effectively supporting learning, what are we arguing about?

The discussion about aesthetics versus effectiveness also needs to include the issue of cost. As my colleague Lane Fischer likes to say, 'with OER there are two ways to win.' What he means is this: in any study comparing the cost of and level of student learning facilitated by commercial materials and OER, OER will always cost less. Therefore, there are only three possible outcomes – OER save money but support poorer learning, OER save money and support the same learning, and OER save money and support better learning. Of the three possible outcomes, two of them favor OER. Getting the same outcomes for less money is obviously a win, and finding better student outcomes for less money is like hitting a walk-off grand slam. This is why 'student success per dollar,' a measure of the percentage of students who receive a C or better final grade against the cost of textbooks required for a course, continues to fascinate me. *When we let commercial publishers dictate the terms of comparison – graphic design, editorial process, peer review – we've already lost. We need to shift the dialog so that OER are judged on the only metric that actually matters – effectiveness. If we can push farther, to measures of effectiveness per dollar spent, we can win.*

Following Open

As I pivoted to this new focus I had the opportunity to partner with Kim Thanos and others on a grant-funded project called Kaleidoscope. This was a Gates-funded project with goals that may sound familiar. We committed to avoid creating new OER and instead reuse existing OER to replace textbooks in classes at eight community colleges around the United States. We learned again that helping faculty think differently about where they find teaching and learning resources, how they select them, and how they use them to support learning is significantly more complicated than it might seem at first. Again, no single repository or tool could catalyze the degree of change we hoped to see. Of course, making it easier rather than harder to find and remix OER aided our cause, but the main determinant of the success of Kaleidoscope was the hundreds of hours of encouragement, training, and support provided by the core team to the faculty partners (together with a willingness on the part of the core team to be yelled and cried at in frustration from time to time by faculty.)

Teaching, it turns out, is a deeply human endeavor. Those humans who teach have a wide range of deeply entrenched and conflicting habits, biases, incentives, and values. Building your WhizBang app with Twitter Bootstrap or using the IEEE LOM metadata standard isn't going to overcome them. From Kaleidoscope we learned, yet again, that large-scale change is best (and perhaps only) accomplished by good old-fashioned handholding, support, and encouragement. More on our process for doing this below.

As Kaleidoscope ended, we began the process of applying for a renewal grant for a second phase of Kaleidoscope in which we would add 20 more institutional partners. I began to appreciate just how much we had learned about OER adoption. We had learned a lot about how to do it wrong, and something about how to do it effectively. Originally I had thought that when the grant funding ended I would turn my full focus back to my tenured faculty position, and that Kim and the others would do likewise. But I began to realize that if we all went back to our previous jobs that learning would go with us. I can remember specifically asking myself, 'What? You're going to write an article about everything you learned about OER adoption, publish it, and then someone *else* is going to quit *their* job and go apply all the lessons you learned to move OER adoption forward around the country?'

The unthinkable started happening before I even realized it – I seriously began to consider stepping away, at least temporarily, from academia.

Then several things happened at once. The Gates Foundation approved our request for renewal funding, and Kim and I founded Lumen Learning as the entity that would carry this important work forward. But I was still on the fence about what to do personally. The most sensible path forward was to remain full-time at BYU and have Lumen buy out a portion of my time. After all, not only was I tenured at BYU, with a matching contribution to my retirement each month from the university, but BYU is one of the few institutions in the US that also has a pension plan. Leaving BYU would mean walking away from an incredibly secure future for my family. However, the more I talked with Kim about the kind of change we thought we could create in the world, the more my wife Elaine and I felt that I had a responsibility – a calling, or a sacred obligation – to keep pushing forward my work on OER.

We decided I would apply for a year of 'leave without pay' from BYU for calendar 2013 and see what we could make happen in that period. Almost immediately after making this decision, I was awarded a fellowship by the Shuttleworth Foundation. The Shuttleworth Fellowship would replace my salary for a year on the condition that I focus my full attention on supporting OER adoption. It was an incredible, timely confirmation that we had made the right choice.

Designing Open

Since we founded Lumen Learning in October 2012 I've learned even more about OER adoption. We've worked our way through successes and failures to a very straightforward model of supporting faculty in making the move from commercial textbooks to OER. The first lesson remains and will remain – systemic change requires dedicated, ongoing support from people who care. I believe the second lesson is wrapped up in the question 'how much instructional design value can we realize in the shortest amount of time and effort from faculty?'

Any seasoned instructional designer will tell you that the overwhelming majority of faculty feel like their terminal degree in their discipline is enough to make them a decent teacher. Even suggesting to most faculty members that their instruction could be improved is seen as insulting. However, once every decade or so a major change comes along that sends faculty looking for help. For example, the opportunity to teach online using a learning management system will send many first-timers looking for support from their Center for Teaching and Learning. In the hands of a skilled instructional designer, the help provided won't end with 'point here, click there,' but will include course redesign work that significantly improves the effectiveness of the course. The improvements are not characterized as strengthening weaknesses in the faculty's current practice, but as new affordances offered by new technologies the faculty member can now leverage for their students' benefit. These infrequent and narrow windows of time are, generally speaking, the only times when faculty are open to significantly improving their courses.

When handled adeptly, the move from traditional materials to OER creates a window of opportunity to improve the quality of teaching and learning. (While adopting an open textbook in place of a commercial textbook saves students money and is generally a good thing, a straight across swap of this kind does not create such a window.) Lumen's model for working with faculty leverages the novelty of OER to covertly introduce faculty to a range of basic instructional design principles.

For example, no principle of instructional design is more basic than assuring that the stated goals of a course match what you're actually assessing, and that these both match what students are actually reading about and talking about in class. Instructional designers refer to this as 'alignment,' and the general principle is that course learning outcomes should be directly aligned with assessments, which should in turn be directly aligned with readings, videos, discussions, and other activities.

Many faculty believe that the first step of using OER in place of commercial textbooks is to find suitable OER. This is one of Lumen's primary instructional design attack vectors. We scaffold the OER selection process for faculty by providing them with a spreadsheet (which I will slightly oversimplify here) in which they list their course learning outcomes in Column A. We then provide them with several previously license-vetted collections of OER related to their course, and encourage them to select one or more OER they feel will best support student mastery of each outcome. The link(s) to these OER go in Column B, so that they remain visually aligned straight across from the course learning outcomes they support. Finally, a description of the assessments appropriate for each outcome goes in Column C, again creating a clean horizontal alignment from each course learning outcome, to the content students will use to study it, to the assessment they will take to demonstrate they have mastered it.

This simple process, one of several we do with faculty, can usually be completed in a one-day workshop, but creates a wide range of benefits to teaching and learning. For example, faculty frequently realize that their course learning outcomes are underdeveloped, and they strengthen, clarify, and add to them. This is a significant professional development activity in and of itself. Faculty also frequently realize that they're covering much more content in their class than they ever intended to, and make comments like, 'I guess I thought if I'm going to make students buy a US$160 book, I wanted them to feel like they're getting their money's worth, so I covered every chapter.' Eliminating these less important topics provides faculty with extra time to cover the topics the course is actually supposed to focus on in more depth. Perhaps most telling of all, students whose faculty go through this process often provide feedback on end of semester course evaluations along the lines of 'I loved that the things we discussed in class were actually related to the readings we did before class!' *The teaching and learning benefits of this kind of small structural improvement are so powerful and obvious that students will mention the difference in an open-ended comment box, unprompted.*

Defending Open

'Serious' instructional designers and learning scientists will no doubt complain that Lumen Learning's simplified approaches to working with faculty gloss over the subtlety and nuance of their fields, and as an instructional designer I fully agree. The approach we have evolved at Lumen is not one that tries to give every faculty member a graduate degree in learning sciences, rather it is a ruthlessly pragmatic approach that asks 'how much can we improve student learning during our interactions with these faculty? What are the highest impact, lowest effort things that faculty probably aren't doing, and how can we integrate them into the OER adoption process?'

This is the point where academics who venture outside the Ivory Tower are typically attacked by their peers. 'How dare you defile the purity of our discipline! How dare you water it down for popular consumption!' One of my favorite sayings – I wish I knew you said it – is 'In theory, there's no difference between theory and practice. But in practice, there is.' This is nowhere more obvious than in trying to support the broad adoption of OER together with effective teaching and learning practices. The bridge from efficiently espousing theories in the classroom to effectively supporting their implementation in the world at times feels like a rope bridge across a great chasm – tenuous, swaying with every gust of wind. But I must admit that building this bridge again and again, and helping people cross it, is some of the most exciting and rewarding work I have ever done.

I have become an outspoken advocate for the idea that academics need to engage more directly in real world work, and do it in collaboration with their students whenever possible. I was significantly emboldened in my thinking and speaking out on this topic by Tom Reeves, whose work on socially responsible research continues to be an inspiration to me. He challenges educational researchers to stop focusing our research on things (e.g., learning analytics or 3D printing) and instead to start focusing our research on problems (e.g., poverty or illiteracy). As I continue to engage in research with my colleagues in the Open Education Group, including John Hilton and Lane Fischer, we fight to maintain this perspective and not let our research devolve into inert studies of OER. *Those working in open education, whether as advocates, creators, teachers, researchers, or in other capacities, would do well to continually focus and refocus their efforts on solving specific problems.*

Growing Open

At the beginning of the chapter I said that I embarked on a journey in a general direction but without a specific destination. Almost 20 years later, I can now see the specific place I am hoping we arrive in the future.

To understand the future of open we must first understand the present of open. In as much as open means free plus permissions, the primary function of open is to create opportunities and expand potential. Consequently, I believe that all meaningful activity in the future of open will fall into one of two categories: further expanding educational opportunities and potential by means of open, and directly supporting people in actively taking advantage of these additional opportunities and potential. You might say that the future of open is about simultaneously increasing negative liberty and positive liberty for teachers and learners.

In order to further expand educational opportunities and potential, we must move beyond our current, narrow conceptions of OER (read: textbooks with open licenses) to a more expansive view that includes all of the core pieces of the intellectual infrastructure of education. When each of these components is opened, I refer to the collection as the Open Education Infrastructure:

- Open Competencies.
- Open Educational Resources.
- Open Assessments.
- Open Credentials.

The full stack must be open because there are critical interdependencies between the components. Until the full stack of our intellectual infrastructure becomes open, truly democratized innovation and permissionless innovation will be impossible.

Open Competencies exist, but they exist primarily as isolated bullets in scattered openly licensed syllabi. These competencies need to be harvested, synthesized, and mapped together in order to create the disciplinary equivalent of Google Maps (actually, OpenStreetMaps would be a better metaphor). In Introduction to Psychology, for example, what are the primary topics? How do they relate to each other? What are the prerequisite relationships between them? What are their relative difficulties? Annotating these Open Competencies using aggregate data from student interactions with OER and student results on Open Assessments will provide us with empirically validated maps of (and myriad pathways through) the disciplines. OER, Open Assessments, and Open Credentials can then be aligned with Open Competencies.

Open Educational Resources are, of course, the most pervasive and best understood component of the Open Education Infrastructure. However, we will need to move beyond the idea that OER are a textbook substitute. The textbook metaphor carries too much conceptual baggage with faculty for the metaphor to be useful in the long term.

Substantive intellectual and practical work remains to be done on Open Assessments. First, questions must be answered regarding the integrity and security of assessments that are openly licensed. Second, as students and faculty (neither of whom are trained in creating valid, reliable assessments) create and contribute a wide range of Open Assessments to the community, we will need to develop techniques for evaluating and improving assessments on the ground and contributing these improvements back to the community. The assessment pilot testing methods used by companies like Educational Testing Service (ETS) may serve as inspiration here.

Open Credentials are certifications that learners own completely and can reuse and redistribute without involving a third party like the college registrar. They may be awarded at the level of traditional degrees or may be aligned with specific Open Competencies and awarded at the individual competency level. These credentials can be regrouped and remixed by learners to highlight different aspects of the learner's expertise, depending on the context in which they are presenting themselves. Each credential must be tamper-proof so that those who evaluate the Open Credential can validate and trust its origin. Mozilla's work on the Open Badges Infrastructure has demonstrated one method of awarding an Open Credential to a learner. While much of the intellectual work on Open Credentials has come a long way, it has recently stagnated because of the public's distaste for the 'badge' branding.

As we engage in the unglamorous, workaday slog of laying the rails and paving the roads of the Open Education Infrastructure, we also need to provide teachers and learners with help. Specifically, they need help understanding what new opportunities now exist and they need to see positive, relatable examples of people like themselves leveraging this new potential in their own classrooms and online courses. Yes, new tools will be important (the freight trains, cars, and long-haul trucks to run on the rails and roads), but these tools will be of absolutely no use if we do not provide significant, proactive support to faculty and learners that teaches them how to use them.

As I have repeated over and over again, we are engaged in a systemic change process – a *human* change process. Massive changes like those we hope to enable by building out the Open Education Infrastructure begin with small steps, like helping faculty create, share, and adopt OER. These steps must be carefully supported and encouraged by people who are committed to their immediate success and who have the long-term vision of what education can become.

References

AECTx: David Wiley: You Have SuperPowers. [Video file]. Retrieved from https://www.youtube.com/watch?v=DuBBhjMnzcg

Bootstrap. (2016). *Main page.* Retrieved from http://getbootstrap.com/

Creative Commons. (2016). *About the licenses.* Retrieved from https://creative commons.org/licenses/

Lumen Learning. (2014). *Home page.* Retrieved from http://lumenlearning. com/

MIT OpenCourseWare. (2006). *2005 Program evaluation findings report.* Retrieved from http://ocw.mit.edu/ans7870/global/05_Prog_Eval_Report_Final.pdf

Mountain Heights Academy. (2016). *Welcome to Mountain Heights Academy.* Retrieved from http://www.mountainheightsacademy.org/

Next Generation Learning Challenges. (2015). *Cerritos college with Lumen Learning.* Retrieved from http://nextgenlearning.org/grantee/cerritos-college-lumen-learning

Open Content. (1999). *Open publication license.* Retrieved from http://open content.org/openpub/

Open Education Group. (2016). *Welcome.* Retrieved from http://openedgroup. org/

Open Source Initiative. (2012). *History of the OSI.* Retrieved from https://open-source.org/history

OpenStreetMap. (2016). *Welcome to OpenStreetMap!* Retrieved from http://www.openstreetmap.org/

Price, J. L. (2012). *Textbook bling: An evaluation of textbook quality and usability in open educational resources versus traditionally published textbooks.* (2012). All Theses and Dissertations. Paper 3327. Retrieved from http://scholarsarchive.byu.edu/cgi/viewcontent.cgi?article=4326&context=etd

Reeves, T.C. (2016). *Socially responsible educational technology research.* Retrieved from https://www.academia.edu/19294287/Socially_Responsible_Educational_Technology_Research

Shuttleworth Foundation. (2016). *Fellowship.* Retrieved from https://shuttleworthfoundation.org/fellows/

Slashdot. (2016). *Stories.* Retrieved from https://slashdot.org/

SourceForge. (2016). *eduCommons.* Retrieved from https://sourceforge.net/projects/educommons/

Thierer, A. (2014). *Embracing a culture of permissionless innovation.* Retrieved from http://www.cato.org/publications/cato-online-forum/embracing-culture-permissionless-innovation

von Hippel, E. (2005). *Democratizing innovation.* Cambridge, MA: The MIT Press. Retrieved from http://web.mit.edu/evhippel/www/democ1.htm

Web Archive. (1999). *OpenContent.* Retrieved from https://web.archive.org/web/19990117060918/http://www.opencontent.org/home.shtml

Web Archive. (2006). *The center for open sustainable learning.* Retrieved from https://web.archive.org/web/20060712031919/http:/cosl.usu.edu/

Wikipedia. (2016a). *Rivalry (economics).* Retrieved from https://en.wikipedia.org/wiki/Rivalry_(economics)

Wikipedia. (2016b). *Learning object metadata.* Retrieved from https://en.wikipedia.org/wiki/Learning_object_metadata

Wikipedia. (2016c). *Negative liberty.* Retrieved from https://en.wikipedia.org/wiki/Negative_liberty

Wikipedia. (2016d). *Positive liberty.* Retrieved from https://en.wikipedia.org/wiki/Positive_liberty

Wiley, D. (2016, January 29). The Consensus around "Open". [Web log comment]. Retrieved from http://opencontent.org/blog/archives/4397

Wiley, D., Hilton III, J. L., Ellington, S., & Hall, T. (2012). A preliminary examination of the cost savings and learning impacts of using open textbooks in middle and high school science classes. *The International Review Of Research In Open And Distributed Learning, 13*(3), 262–276. Retrieved from http://www.irrodl.org/index.php/irrodl/article/view/1153/2256

Wiley, D. (2014, July 15). The Open Education Infrastructure, and Why We Must Build It. [Web log comment]. Retrieved from http://opencontent.org/blog/archives/3410

Open-Source for Educational Materials Making Textbooks Cheaper and Better

Ed Diener[*], Carol Diener[†] and Robert Biswas-Diener[‡]

[*]University of Virginia

[†]University of Illinois

[‡]Noba, robert@nobaproject.com

Editors' Commentary

Aside from being expensive, traditional textbooks are also rigidly structured, environmentally unfriendly, and unnecessarily long. As a result, although most open textbook projects attempt to produce an openly licensed version of a resource that most faculty will find familiar, authors Ed Diener, Carol Diener, and Robert Biswas-Diener—all affiliated with Noba—set out to do something different. The model that they have created, with leading scholars authoring over 100 brief modules (supported with high quality ancillary resources) that faculty may customize using a user-friendly interface, represents a new and innovative model for open educators in other disciplines. In this chapter the authors discuss their rationale for the Noba project, describe some of its successes and challenges, and share their hopes for the future of open education.

When we reflected on our combined decades working as instructors in academia, we saw strengths in the educational system in which we worked; but also shortcomings. One of the biggest problems that emerged across our tenure as instructors was the rising cost of a college education. There are many reasons

How to cite this book chapter:

Diener, E, Diener, C and Biswas-Diener, R. 2017. Open-Source for Educational Materials Making Textbooks Cheaper and Better. In: Jhangiani, R S and Biswas-Diener, R. (eds.) *Open: The Philosophy and Practices that are Revolutionizing Education and Science.* Pp. 209–217. London: Ubiquity Press. DOI: https://doi.org/10.5334/bbc.p. License: CC-BY 4.0

why education has become more expensive but one factor that stood out to us was the cost to students of textbooks. We landed on this factor because—unlike cost of living or tuition—it is one that faculty have a direct influence upon. According to The Enterprise Institute[1] textbook prices have increased more than 800 per cent since 1978. By contrast, the Consumer Price Index—a measure of variation in the price of common retail goods—has only risen 250% in that same period. The consequences of this are predictable: A US PIRG survey[2] of more than 2,000 students from 156 campuses revealed that 65% of the students surveyed indicated that they had decided against buying a required course text because of the expense. Over 90% do so knowing that it could adversely affect their grades. Students Surveys reveal that many students never buy the textbook for a class, and just try to get along without it. In this chapter, we will discuss the problematic landscape of textbooks and relate the story of how we came to address these problems through Noba.

Traditional Textbooks

Textbook prices are kept high by several means:

1. **New Editions**. First, publishers ask authors to update their books frequently, meaning that older and used books cannot be assigned by professors because they are viewed as 'out of date.' Whether the scholarship in various fields that undergraduates need to learn has actually changed enough to warrant continual revisions of textbooks is, in our view, doubtful. One clear example of this can be seen in the instance of the release of the latest Diagnostic and Statistical Manual of Mental Disorders.[3] Following the publication of the DSM-5 many textbook publishers rushed new introductory psychology books to market that promised special DSM-5 updates. It is not entirely clear, however, that the taxonomic structure or diagnostic criteria for clinical professionals found in the DSM are an important enough influence on introductory undergraduate and high school psychology that a revised book is warranted. While experts might argue over the scale of changes between the DSM-IV and DSM-5 (were they large or small?) there were very few changes that would affect how basic material was introduced to general psychology students. It is possible that the new editions of DSM-5 compatible textbooks reflect a market demand but it is also possible that this is an illustration of the profit driven edition cycle common in academic publishing.
2. **Non-traditional economic model**. Another factor that keeps textbooks high is that professors select the books, but students pay for them. In our experience this means that instructors really are not price sensitive, and this allows much higher prices than if professors or universities were paying for the books. In a traditional economic model supply and competition would

serve to reduce costs. The clearest corollary with the economic model of textbooks—at least in the United States – is prescription medicine. In the US doctors are the ones who often choose the medication but patients are the ones who pay for them. Not surprisingly, prescription medications, according to Consumer Reports,[4] are also rising at an alarming rate (1 in 10 medications rose in price by more than 100 per cent over a one year period from 2013–2014).

3. **Size.** The third factor that makes textbooks prices so lofty is that they try to cover a huge amount of material in an attempt to cover any and every topic that a professor might want. This results in very, very large texts that cover much more material than any student could possibly learn in a single course. So-called brief textbooks are a notable market solution to this problem and, in themselves, a recognition that traditional texts may be too long.

Beyond the issue of expense, we noticed a number of additional problems with traditional textbooks. For one thing, they use a lot of paper, and many students are increasingly comfortable with digital materials, including texts. Although surveys commonly point to the conclusion that the vast majority of students continue to use print textbooks many also find that as many as 30% prefer digital texts[5] and 80% find e-books helpful in completing assignments and test preparation.[6] Second, textbooks present certain material and present it in a fixed order. Thus, instructors who want to customize their course face a challenge in organizing the materials for students. For decades, instructors have tailored course content to reflect their own preferences and students have endured syllabi that suggest reading 'Chapter 2' and then 'Chapters 4 and 9' and then 'Chapter 6.' Digital texts could easily solve this problem. We also heard professors say that they would like more on certain topics, and coverage of topics that are not in the text they choose. The so-called standard canon—the topics found in many introductory psychology books—appears to preference some historically accepted topics such as hypnosis, while overlooking others such as the current replication crisis, the academic publication process, and knowledge emotions. Again, computer based textbooks offer a solution because they allow professors to add and subtract material from their textbooks. Finally, digital textbooks also allow added features such as embedded learning assignments, videos, interactive simulations and hyperlinks that have the potential for increasing student learning while they read the text.

One final issue related to the cost of traditional textbooks is that many scholarships will cover tuition, but not textbooks. Thus, the students are on their own, and a shocking number of students report not buying the books for all of their classes. This puts needy students at a disadvantage compared to more affluent ones.

Although we see the problems with textbooks we—the authors of this chapter—do not consider ourselves above them. We have assigned traditional textbooks. Instructors buy into the current model of expensive texts sold by commercial publishers for several reasons. One reason is that many of the text-books are well-written and of reasonably high quality. They contain interesting side bars and high quality graphics. They are well edited. Their popularity is also based on other factors. For instance, professors find it convenient and easy to use commercial texts because they come with a variety of teaching aids such as instructors' manuals, test banks, and PowerPoint presentations. Last, traditional textbooks are popular for the same reason that some brands of beer are: namely, they are a commercial product at the tip of a large marketing spear. Sales reps send emails, free samples, and drop by the office to answer questions. Some publishers of traditional texts supply faculty with travel grants and other 'development money.' Somewhere between the high cost, the inflexible structure, and the lack of student input there is a clear problem in need of a solution.

The Advantages of Digital Textbooks

There are several suggestions alluded to in the previous section. All of them are based on the properties of the digital medium. It might sound strange to focus so heavily on digital in a book about open resources but the truth is 'digital' and 'open' go hand in glove. In open education circles advocates often talk about an 'abundance model' rather than the traditional 'scarcity model.' It is the advent of digital technology that allows us to easily create, modify and share materials at an unpresented rate. The proliferation of access to the internet has dovetailed with the gaining traction of the open movement. It is the reason why open exists today and not back when scribes were scratching cuneiform on clay tablets.

Based on the comments above, here is a summary of the advantages of digital (and open-source) texts:

1. **Less expensive**. Publication costs do not include paper, ink, packaging or shipping. Revisions can be made more easily without rewriting the entire book, and the costs of shipping and printing are negligible. Therefore the notion of 'revised editions' can be supplanted by 'revisions as necessary.' Even if students choose to print out a hard-copy of their open textbook, or use a print on demand service the costs are much lower than traditional texts. For example, Noba's textbook 'Discover Psychology 2.0' currently retails for less than 15 US dollars plus shipping. Admittedly, it is a softcover, black and white book, but we regularly receive student feedback suggesting this is an acceptable trade.

2. **Easier on the environment**. Digital texts do not require physical manufacture or shipping.

3. **Foreign access**. It is much easier to access materials, especially in poor nations where costs are a serious concern. Many economically developing nations have invested in internet infrastructure.

4. **Individualization**. Instructors can select only the modules he or she wants to cover in a course, and can add or subtract material from chapters. An extension of the idea of individualization is localization. The open format allows instructors around the world to modify content so that it is more relevant and relatable to their students.

5. **Alternative chapters can be made available.** Digital textbooks can use alternate forms of the same material. For example, a text might offer an overview chapter on 'The Brain and Nervous System' but also offer more specialized units on 'Neurons' and 'The Biochemistry of Love.'

6. **Active learning**. The digital format offers a variety of methods for increasing in the moment learning including "mouse overs" (e.g., definitions pop up on screen), 'adaptive learning' (tests embedded in the chapter that track individual student performance and adapt to it), and hyperlinks to related videos and readings.

7. **Accessibility**. The open format has fewer restrictions regarding printing, copy and pasting, sharing, and re-sizing—to name a few—which means that these resources support assistive learning for students who need accommodation.

The Noba Model

Two of us (ED and CD) decided to fund Noba because of the problems we saw with the current textbook model. We felt comfortable focusing our efforts exclusively on psychology because it is considered a high enrollment course with a large potential for student savings. Our initial idea was to have experts for each topic in psychology write chapters on their respective areas. For example, we received our chapter on 'Eyewitness testimony' from Elizabeth Loftus and our chapter on 'Evolutionary theories in psychology' from David Buss. By focusing exclusively on our own area of expertise (psychology), and by investing in a model that emphasizes expert created content we have largely been able to side step some of the common skepticism of quality faced by many OERs. In fact, we have developed a sophisticated quality assurance program that includes:

1. Expert created content.
2. Editorial review.
3. Peer review.
4. Student review.
5. International review.
6. Accessibility review.
7. Empirical review.

We also opted for a modular approach in which instructors could drag and drop each chapter in which they were interested into a unique course textbook and assign it only to their own students. We created 100 different modules ranging from 'research methods' to 'creativity' to 'mood disorders' covering all corners of psychology. The modular approach means that Noba materials can be used either as a collated textbook or as supplemental reading. Of course, all of these materials are Creative Commons licensed and it was our strong intention to make them free for all students.

To generate enthusiasm for Noba we also directed money toward faculty and students in the form of grants and awards, none of which were contingent on using Noba. On the faculty side we awarded grants for the review of our materials and for experimenting with new digital capabilities. For example, because textbooks can now be individualized it is possible that instructors could assign core content and each individual student could select additional chapters that represented his or her own unique interest. It is possible that having 30 unique textbooks in a class of 30 students is a management nightmare for instructors. It is equally possible that this method is an unprecedented means of engaging students and increasing interest. We'll have to wait until the data are in to see. On the student side we awarded 10 thousand dollars each of two years for the creation of short videos that would become a permanent part of Noba catalog.

It is here, perhaps, that we are most proud of Noba. Earlier digital and open source textbooks were simply electronic versions of traditional textbooks. Because the Noba system is built around modules it is changing the way we think about what a textbook is, what it can do, and how it can impact learning.

Then, an interesting thing happened. Instructors did not flock to our site. Many seemed unmoved by the free price tag. Some refused to mention Noba to their local psychology clubs. Others criticized our experts as being 'biased.' We heard first hand reports that sales reps from large publishers were warning faculty members away from open source materials such as Noba, suggesting that they are low quality. It was surprising and unsettling. Ultimately, we interviewed instructors and learned a simple truth: they are dedicated and passionate teachers who are often overworked and who have little reserve energy to evaluate a new text, let alone create new lectures and tests if they adopt one.

Based on this insight we shifted our priorities from textbook creation to instructor support. We collaborated with a large team of experts in both psychology and instructional design to create high quality test banks, instructor manuals and PowerPoint presentations. We created a print on demand version of our textbooks that sells for less than US$15. We partnered with Cerego, an adaptive learning technology company, to embed each of our modules with an adaptive quiz function. We revised many modules so that they employed simpler language and more international examples. We created a blog of practical teaching ideas and offer essays on a variety of teaching topics ranging from teaching large classes to teaching biological psychology. We gave additional grants to instructors to pilot a review of our materials and create rubrics for evaluating Noba that can be used by individual instructors and departments both. We ensured that our materials were accessible to those with visual impairments and other disabilities. In essence, we tried to think of every small concern or headache an instructor might face in deciding whether to adopt Noba and deal with it. And, if it was not clear before, it's all free to all of our users.

The result of our efforts is—three years in—visible. We are being officially adopted at two and four year institutions such as Northwest Vista College (a 2-year institution using Noba with approximately 1,500 students a year), West Virginia University (approximately 3,000 students a year), and East Georgia State University (an institution that reports US$114 per student in savings.[7] We are being recommended by our colleagues and are receiving higher satisfaction and engagement ratings from our users. Most importantly, we have saved students—using a conservative estimate—more than two million dollars in textbook fees.

We also track out impact qualitatively through feedback be receive from instructors and departments that use Noba. Here are two such examples that highlight both the financial and pedagogic benefits:

> 'Teaching with Noba gives me freedom as an instructor. It gives me freedom to assign as little or as much to my students as I would like. It gives me freedom to supplement my teaching rather than overshadow it. It gives me the freedom of knowing my students (some of the poorest in the nation) are not falling behind because they are waiting on financial aid to come through.' –Raechel Soicher, Linn-Benton Community College
>
> 'I am thrilled that we have such a wonderful resource for our students. We have over 3,000 Introduction to Psychology students each year at WVU and you can imagine how much money this saves our students. Even after the semester is in its 2nd week we have students asking where to buy the textbook. It's nice to remind them that the resource is free.' – Constance Toffle, West Virginia University

The Future of Open Education

The cost of college education has been spiraling upward, and textbooks are one part of this trend. Textbooks are much more expensive than they need to be, and open-source texts are one solution to the problem. Furthermore, the digital format has the potential to radically transform the way that texts are used as instructional aids and we are only beginning to scratch the surface of this potential. In the future, instructors will better be able to modify the actual text, the examples, and the featured research contained within modules. To some degree, that very concept is anathema to academic culture. Many professors place a premium on their intellectual property and value their own authorial voice. The notion that someone else might tinker with that can—admittedly—be unsettling. Rather than simply dismissing the idea of open-source collaboration future educators must engage in a discussion about how this process can occur while still preserving the scientific rigor we all prize so highly.

It is our hope that in the near future we will see a sea change in the attitudes of individuals and institutions regarding open materials. We would like to see a day where high quality open materials are the standard and students must opt out to purchase additional resources. Boards of trustees, university administration, alumni groups, student groups, and even legislatures can bring pressure to bear to use cheaper textbook alternatives. Many are fearful of seeming to tread on academic freedom, but selecting from among the less expensive open-source textbooks does not truly limit in any way the material they cover in their courses—and this is what academic freedom is.

Notes

[1] Weissmann, 2013.
[2] Senack, 2014.
[3] American Psychiatric Association, 2013.
[4] Consumer Reports, 2015.
[5] Bolkan, 2015.
[6] Falc, 2013.
[7] Affordable Learning Georgia, 2016.

References

Affordable Learning Georgia. (2016). Textbook transformation grants: Round one projects gallery fall 2014 – spring 2015. Retrieved from http://www.affordablelearninggeorgia.org/site/round1

American Psychiatric Association. (2013). Diagnostic and statistical manual of mental disorders (5th ed.). Washington, DC: Author.

Bolkan, J. (2015). Survey: Most students prefer traditional texts over e-books. Retrieved from https://campustechnology.com/articles/2015/09/01/survey-most-students-prefer-traditional-texts-over-ebooks.aspx

Consumer Reports. (2015). Are you paying more for your Rx meds? A consumer reports' poll shows one-third of Americans hit by high drug prices. Retrieved from http://www.consumerreports.org/cro/news/2015/08/are-you-paying-more-for-your-meds/index.htm

Falc, E. O. (2013). An assessment of college students' attitudes towards using an online e-textbook. *Interdisciplinary Journal of E-Learning and Learning Objects*, 9, 1–12.

Senack, E. (2014). Fixing the broken textbook market: How students respond to high textbook costs and demand alternatives. Retrieved from http://

uspirg.org/sites/pirg/files/reports/NATIONAL%20Fixing%20Broken%20 Textbooks%20Report1.pdf

Weissmann, J. (2013, January 3). Why are college textbooks so absurdly expensive? The Atlantic. Retrieved from http://www.theatlantic.com/ business/archive/2013/01/why-are-college-textbooks-so-absurdly-expen- sive/266801/

Free is Not Enough

Richard Baraniuk[*], Nicole Finkbeiner[†], David Harris[†],
Dani Nicholson[†] and Daniel Williamson[†]

[*]Rice University & OpenStax, richb@rice.edu

[†]OpenStax

Editors' Commentary

It is difficult to talk about open education without focusing on open education resources. Among the resources it is hard not to mention textbooks. Among textbooks, it is impossible not to discuss OpenStax. In this chapter, the authors—all of whom are affiliated with what is perhaps the best known open textbook publisher in the world—tell the story of OpenStax. In doing so they argue—as the name of the chapter implies—that merely being free is not enough to justify openness. They argue that high quality, course relevance, and instructor support materials are key elements of effective open education. They share the lessons they have learned at OpenStax regarding economic sustainability of OERs.

Introduction

OpenStax CNX, then Connexions, was founded in 1999 with three primary goals: (1) to convey the interconnected nature of knowledge across disciplines, courses, and curricula; (2) to move away from a solitary authoring, publishing, and learning process to one based on connecting people in open, global learning communities that share knowledge; and (3) to support personalized learning. OpenStax CNX has grown into one of the largest and most used

---

How to cite this book chapter:

Baraniuk, R, Finkbeiner, N, Harris, D, Nicholson, D and Williamson, D. 2017. Free is Not Enough. In: Jhangiani, R S and Biswas-Diener, R. (eds.) *Open: The Philosophy and Practices that are Revolutionizing Education and Science.* Pp. 219–226. London: Ubiquity Press. DOI: https://doi.org/10.5334/bbc.q. License: CC-BY 4.0

OER platforms – each month millions of users access over 20,000 educational 'building blocks' and thousands of e-textbooks. In addition to web and e-book outputs, a sophisticated print-on-demand system enables the production of inexpensive paper books for those who prefer or need them, at a fraction of the cost of books from a conventional publisher.

This approach was widely hailed throughout the open education community, and other groups such as Merlot, OER Commons, and Orange Grove have followed a similar mission. Hundreds of thousands of learning objects were created, and these have been used by millions of learners. Mission accomplished – at least, that's what we thought.

By 2008, it became clear that simply providing a delivery platform for course materials was not enough to increase access for the majority of students. Faculty who had the time, experience, and drive to make educational resources created these course materials. Although these resources reached thousands of students and no doubt improved access for many, it didn't provide the widespread shift to access for all that OpenStax and others in the open community envisioned. This observation, coupled with the increasingly pervasive issue of student debt, forced the team to take a close look at our platform and ask questions about scalability, sustainability, and the future of access to educational content.

Rethinking the doctrine

For the past several years, faculty have been asked to do more with less, their time stretched further each year. OpenStax found that the faculty taking advantage of OpenStax CNX were not the faculty who, in many cases, most needed open and easily accessible resources. The faculty that used OpenStax CNX the most were faculty who had time in their teaching schedules and were personally motivated to create resources and add them to the library. We needed to find a way to efficiently serve the majority of faculty – the faculty who do not have time to create educational resources or piece together quality resources from multiple sources. A good example is the adjunct instructor notified a few weeks – or days – before the start of the term that they will be teaching a course. We needed to serve this majority group of faculty so that we could reach our ultimate goal of improving access to education for all students while encouraging academic freedom. Free resources would not be useful to students if they didn't meet the requirements established by the instructor.

Another factor we considered was that in highly enrolled courses, like psychology or college algebra, the curricula is well-defined and often doesn't vary from school to school. Bearing this in mind, instead of expecting faculty to adapt and create resources, then adopt those resources, we shifted our tenet: adopt developed, high-quality resources, then adapt. By changing our assumptions, we were forced to reconsider the very nature of traditional open educational resources.

OER 2.0

Once OpenStax developed a better understanding of faculty needs, we got to work addressing the shortcomings of the OER concept as it was then defined. With the end goal of helping as many students with the cost of textbooks as possible, we found four key deficiencies that we remedied with the release of a new library of free, peer-reviewed, professionally developed textbooks.

1. *Free is not enough. Materials must meet the quality thresholds set by the community.* OpenStax follows a professional development model and we are finding that our resources scale rapidly. Locally produced materials work locally because the author can provide context, in-class explanation, and supplemental problems. There is also a very good chance that local assessment and local content are well aligned. However, it takes teams of professionals, including authors, reviewers, development editors, graphic designers, and assessment experts, to develop resources that can be used at many institutions without extensive adaptation. This professional development model ensures the content meets quality thresholds set by the community of educators. The more complete and easy to use, the more likely a resource will be adopted. In just over three years, millions of learners across thousands of courses have used OpenStax materials. More importantly, a recent survey of 400 users indicated a re-adoption rate of 96.4%. Quality is sticky.

2. *Meet standard scope and sequence requirements.* Faculty have ever-increasing responsibilities and less time to restructure their courses around new materials, combine materials to create their course, or write their own materials. Creating resources that meet standard scope and sequence requirements significantly reduces the barrier for OER adoption because it takes faculty less time to adopt. Also, if faculty want to adapt the materials or add their own content, it is much easier for them when they have professionally produced materials to build upon. This practice also frees faculty to drive pedagogical reform such as inquiry-based approaches and flipped models, thereby enhancing academic freedom.

3. *Improve discoverability.* OpenStax CNX, like other OER repositories, is burdened by a surplus of content types. A 2014 Babson survey revealed that discoverability of complete course OER was a major hurdle in adoption. To improve discoverability, we positioned our peer-reviewed, professionally developed textbooks separately from OpenStax CNX, while still serving these textbooks via OpenStax CNX and making them available in the OpenStax CNX library. Our peer-reviewed textbooks are available at openstaxcollege.org, where users can download a PDF, follow a link to the OpenStax CNX web view, or order a low-cost print option from Amazon or campus bookstores. This positioning has proved immensely

successful. In fact, one week in September 2015 garnered over one million unique visitors to openstaxcollege.org.

4. *Provide essential, additional resources.* Faculty are accustomed to using additional learning resources such as presentation slides, solution manuals, online homework, and courseware to better manage their courses. This is especially true for adjuncts who often have very limited time to prepare for a last-minute course assignment or are asked to teach introductory survey courses that cover many topics beyond their specific expertise. To address this challenge, OpenStax provides these learning resources either directly or at a low cost through for-profit providers. One of the biggest complaints we had heard from faculty is that they don't like being forced to use a particular homework system because it's the only one paired with the book, so we partner with a wide variety of providers to allow the faculty to choose what is best for them and their students.

OpenStax addressed these issues, while retaining the best qualities of OER. OpenStax books are free and licensed under the Creative Commons Attribution International (CC BY) 4.0 license (except for Calculus, which is CC BY-NC-SA), allowing faculty the academic freedom to utilize the materials however they see fit, whether it be incorporating them into videos for a flipped classroom or adopting them as the primary text for the course. These titles are free, openly licensed, and add much-needed value for faculty and students everywhere.

A primary motivation for using OER is that it can help increase access to education for students. By addressing these four deficiencies, we intervened sooner in the OER creation process and made OER more useful and accessible for the majority of faculty as well as students.

Sustainability

Creating a successful and sustainable model required us to evaluate the current market models. The case for positive disruption of the publishing industry is well documented; the current economic model is broken, student access is declining, and the price-to-value ratio of course materials is no longer sustainable. We found that digital rights management (DRM) restrictions and the lack of collaboration among providers in the market were two key factors that would reform the traditional, monolithic publishing model and pave the way for a successful, sustainable OER model.

Reducing the shackles of DRM

Millions of OpenStax users have enthusiastically embraced the significant reduction in price and the lack of DRM restrictions on OpenStax content. The current DRM restrictions for most of the digital content sold today by traditional publishers are simply not in line with the way students acquire, use, and share information. Current DRM restrictions come from a bygone era of 'unidirectional' information flow and often limit access to a period of days, the number of pages that can be printed, the number of devices content can be accessed on, and the sharing of information.

Fortunately, today's learners have grown up on the web. The opportunity to network, access information, and share knowledge is limitless. Open resources remedy the shortcomings of DRM by allowing users permanent, unfettered, and unlimited access across multiple devices and platforms. Anywhere, at any time, in any format, the ability to share content provides a level of freedom that is igniting innovation and lasting change across the market.

We may look back on this early period and conclude that quality OER went mainstream not only because it was free, but because it provided much-needed freedom for students and educators.

A distributed ecosystem model

It is now common practice for hardware producers and software producers to collaborate in an ecosystem to accelerate innovation and increase the use of their products and services. Traditional publishers, however, have consistently bucked this trend by turning inward and creating 'one size fits all' course materials through force of habit.

At OpenStax, we've found that the path to widespread student savings and sustainability also requires an emerging and vibrant ecosystem. Our allies in our ecosystem improve access, efficiency, and quality by:

- optimizing expertise rather than recreating expertise many times over;
- minimizing the cost to onboard faculty and students;
- improving quality and choice for the community, and;
- providing ongoing mission support fees to support the effort.

Optimizing expertise, not recreating expertise many times over

OpenStax is highly efficient in many areas; however, we would definitely be inefficient in developing online courses, in the distribution of non-electronic product, on-campus marketing, and providing in-classroom response services without the help of our ally organizations. There has been a dramatic increase in the number of positive disruptors in the education industry over the last few years, including Sapling Learning, Lrnr, Top Hat, Redshelf, and WebAssign, to name just a few. Even industry stalwarts like NACSCORP, RR Donnelly, and Wiley are looking to reinvent themselves. For example, NACSCORP signed an agreement with OpenStax in 2014 that allows NACSCORP to manage the bookstore distribution of OpenStax textbooks. With support from their campus bookstores, faculty can now more easily adopt OpenStax textbooks, and students are benefitting from even lower print textbook prices.

These organizations recognize that incorporating OER into their products and services not only reduces their time to market (or their ability to retain existing customers), but allows them to make greater investments in tools that enhance learning. By leveraging OpenStax content, they are maximizing efficiency. These partnerships also allow OpenStax to enter markets without having to develop an entirely new set of competencies. A virtuous cycle of efficiency is created.

Minimizing user acquisition costs

A user acquisition cost is the cost required to have a single student in one course use a resource for the first time. (It should be noted that even free resources have an associated user acquisition cost.) These acquisition costs are exceptionally high for traditional publishers; consider the number of marketing managers, sales representatives, and technology support staff that must be employed to drive adoptions. It is imperative that a non-profit reduce these costs to a bare minimum. By working with ecosystem partners, OER providers can instantly gain

- access to marketing and sales organizations;
- collaboration around social media engagement, and;
- customer contact information when permissible by end user agreements.

OpenStax does not have a sales representative, nor do we plan on hiring a sales force. However, our allies do have marketing groups and sales representatives, and we regularly collaborate with these groups on advocacy campaigns. This collaboration has contributed significantly in the market awareness of OpenStax. The cost to OpenStax has been minimal, but the impact has been priceless.

As more and more faculty adopt OpenStax, the tables are beginning to turn, and we are now able to introduce our base of users to our allies, reducing their customer acquisition costs. This lowers the price that students and institutions need to pay for the resources, perpetuating our cycle of efficiency.

Improving choice and quality for the community

Why should faculty be locked into a platform that aligns only to a specific set of course resources? How rapidly can we move beyond one size fits all resources? The ecosystem model spurs choice by allowing the educator to decide which resources best align to their curricular goals. For example, in physics, we have no fewer than four online homework providers. Each provider has unique attributes that work well for different student populations. The ecosystem is also a competitive market, spurring innovation while keeping end user costs to a minimum. In fact, the cost for OpenStax additional resource services is typically 65% less than comparable incumbent solutions. In the future, OpenStax may produce adaptive technologies offered alongside other applications in the burgeoning ecosystem; however, we will never assume a single choice model for the market because a virtuous cycle of competition benefits the community. Choice drives innovation and creativity.

Providing ongoing mission support fees

Content that scales globally requires philanthropic resources. However, it is our responsibility to make sure we maintain our content and sustain operations. Our allies, who are rapidly increasing in number – greater than 40 organizations as of December 2015 – are excellent stewards of the community. Ally organizations realize that OER should be a taking system for students, and, as businesses they embrace the responsibility to give back when they achieve gains from using openly licensed content. If an educator chooses to utilize the high quality and affordable homework, courseware, or other solutions from our allies, part of the proceeds come back to OpenStax in the form of a mission support fee. In fact, mission support fees have already funded the revision of our sociology and economics texts.

Closing Thoughts and Summary

The open community is wrestling with a supply and demand problem: there are not enough open educational resources suitable for the faculty that either already want to use OER or would be open to adopting OER once introduced. This supply and demand problem is a barrier to the ultimate goal of improving access to education for all. The model that OpenStax has adopted is a practical solution and enables widespread use of OER – not just for faculty with the resources to create it.

OpenStax development costs are an excellent example of the efficiency of this new model. Our content development and production costs are approximately 60% to 70% less when compared to the traditional publisher's model. While it is also true that our development and production costs are higher than the initial creation of locally produced materials, this does not take into account the full picture if scaling is factored into overall costs. Our cost-per-user is extremely low, which is crucial because it allows OpenStax to provide significant savings for a vast number of students. Locally developed OER has an important place in education, but it does not scale affordably and therefore make a significant impact on student access.

We are acutely aware that achieving scale is a dynamic effort and that the market is moving away from the traditional texts; however, the community will always demand effective content. Also, the way in which students interact with content is changing. At OpenStax, we are exploring not only ways in which students learn from our resources, but ways our resources can learn about students. We have teams of researchers investigating the most effective ways to integrate machine learning algorithms and to implement principles from cognitive science, such as spacing and retrieval practice, to improve student comprehension and retention of key concepts. For OpenStax, increasing access also carries with it the responsibility to improve students' return on effort so they are more successful in current and future courses. Thankfully, these current efforts are building on the high-quality, openly licensed content that we have already produced.

In summary, these are some of the tenets that have proven effective in creating scalable open content:

- Professionally produced OER that meets standard scope and sequence requirements has proven to scale effectively.
- Educators adopt and then adapt, not vice-versa.
- Any format, anytime, and anywhere drives usage.
- Affordable may need to trump free at scale, because not every resource can be free.
- A distribution ecosystem that reduces DRM and market costs and provides sustainability can spur virtuous cycles of quality, innovation, and affordability.

It is our mission as an organization to increase access to education for all, not just those with resources; this applies to faculty and students alike. 'Access. The Future of Education' is far more than a tagline. OpenStax is comprised of individuals who have different backgrounds and motivations for choosing this work; however, our binding quality is our pragmatic yet passionate approach to making significant gains in improving access. As members of the open community, we must create a future in our lifetimes where access for all is commonplace and a student is limited only by their aspirations, not the cost of their book.

The BC Open Textbook Project

Mary Burgess

BCcampus, mburgess@bccampus.ca

Editors' Commentary

What would happen if government recognized the potential impact—on students and society as a whole—of open textbooks? Over the four years since it was launched the BC Open Textbook Project has greatly surpassed its goals, on a shoestring budget. In this chapter, author Mary Burgess describes how a small team of committed professionals has managed to build a repository of more than 150 open textbooks and foster the adoption of open textbooks at every public institution in the province by raising awareness, maintaining an academic focus, building capacity within institutions and specific disciplines, connecting people with resources and expertise, and drawing on the knowledge of those who have trodden similar paths. The chapter concludes with a commentary on some of the challenges of quantifying success in this arena and on the role of policy in supporting the next phase of the project.

What is the project

It was October 2012 and about 300 international OER advocates, researchers and practitioners were gathered in the auditorium at UBC Robson Square in Vancouver. John Yap, the BC Minister of Advanced Education, took the podium and announced that the government of British Columbia would provide a $1 million dollar grant to BCcampus to manage a provincial Open Textbook program.

How to cite this book chapter:
Burgess, M. 2017. The BC Open Textbook Project. In: Jhangiani, R S and Biswas-Diener, R. (eds.) *Open: The Philosophy and Practices that are Revolutionizing Education and Science.* Pp. 227–236. London: Ubiquity Press. DOI: https://doi.org/10.5334/bbc.r. License: CC-BY 4.0

A year later, the Ministry announced a further $1 million dollars in funding to produce 20 books for areas in which, as a province, we have skills gaps or projected skills gaps; trades, tourism, technology, healthcare, and adult basic education.

Work on the project has brought both successes and challenges. BCcampus, a government funded agency working in support of BC's post-secondary sector in the areas of teaching, learning, educational technology and open education, is tasked with managing the project on behalf of the Ministry of Advanced Education. The primary goal of the project is to provide flexible and affordable access to higher education in BC by making available 40 openly-licensed textbooks aligned with the most highly enrolled first and second year subject areas in BC public post-secondary institutions. This includes of course, both first and second year Psychology, given the popularity of the discipline with students. We are also striving to use openness as a platform for a shift to more student centred, outcomes based design of instruction that creates meaningful, applied and engaging learning experiences for students. We are focused on capitalizing on the teaching and learning benefits achieved through a culture of sharing.

At the time of writing this chapter, the project is now 3 years old, and we have learned a whole lot about what it takes to make a project of this scale happen.

How we got here

BCcampus has long been advocating for and actively working in the domain of OER. From 2003 to 2012, the organization administered the Online Program Development Fund (OPDF). This fund, provided by the BC Ministry of Advanced Education, enabled faculty and staff in BC institutions to create OER, which ranged from small multimedia elements to full online programs, all of which were licensed openly. Each year, institutions applied for funding and were required to collaborate with each other in order to secure their grant. Grants were decided on by a multi-institutional panel who looked at the proposals with a view to the entire BC system rather than simply what would benefit a single institution. This early work in OER licensing education and in building a culture of collaboration between institutions laid the groundwork for the Open Textbook Project. Like many, I have always seen collaboration and Openness as natural allies. Not only did the OPDF enable efficiency from the perspective that within a single funding system, institutions were able to use each others' resources rather than each one paying for their own, it also demonstrated the true power of Open. When we enable collaboration, we enable a better end result for students. Having many experts co-develop a learning resource and then allowing other experts to revise and improve it leads to resources that are tailored to enable learning against specific outcomes. There is a continuous improvement cycle when many people are allowed to be part of that process that just cannot be replicated when access is restricted.

When the Open Textbook Project was announced, we were very excited by the prospect of improving access to higher education by reducing costs for students. In addition, providing faculty the ability to control the resources they use by adapting them to their own contexts was seen as a major benefit to the learning process.

At the time of this writing, we have 163 textbooks, 703 courses into which our resources have been adopted, and have saved students more than two million dollars. How on earth did we get to this point?

Get out there

Getting input and talking with instructors, senior administrators and other institutional staff as well as students has been a key part of our work since the very beginning of the project. Ultimately we want to ensure this project is owned by the BC post-secondary system, not by BCcampus or by government despite the fact that we see ourselves as having a long term role in supporting this work. Encouraging that sense of ownership means getting out there and talking to people about their challenges, needs and vision. Three months after the project was announced, we held our first meeting of the BC Open Textbook Project Subcommittee. The committee was formed through a process of application, and those applications were vetted to ensure a desire to move open education forward in BC, representation of institution type (research focussed university, teaching focussed university, college and institute), region, and institutional role. We brought together faculty, students, instructional designers, librarians, senior administrators, and bookstore staff. The committee was intended to both inform the project and act as communication agents within their own institutions. Their input was integral to planning the project and to reflecting the system back to us as we put the pieces in place. They weighed in on everything from the best times in the academic year to do calls for proposals, to how much funding would be needed to incentivise faculty to participate. For example, we were advised to work closely with librarians as advocates. We took that very seriously and we now have a very active group of librarians in BC called BCOER who are collaborating with each other to support the adoption of open resources on their own campuses. The committee members acted as institutional contacts, advocates, and conduits to more communications.

Like our OER colleagues, BCcampus staff have given many presentations and workshops over the last 3 years. Being present within institutions to hear the questions and develop a deeper understanding about faculty barriers to adoption was extremely helpful for our processes. We did presentations to articulation committees (discipline panels), at faculty meetings, library meetings and anywhere else we could get an invitation. Early on, we asked if we could come. We are, thankfully, now at a point where we are asked to come. The questions we are asked at these sessions have also changed. When the project was in its infancy, many of the questions were related to quality. Specifically, there was misinformation about the quality of OER and whether it was possible to get a high quality end result when a traditional textbook publishing model was not used. This concern influenced our work to ensure quality was a focus for us, and brought a more academic focus to the project that I will discuss later in the chapter. Now when we attend these events, it is much more likely that we will be asked to help someone find an open textbook that specifically meets their needs, or how to do an adaptation of an existing open textbook. This shift in attitude illustrates the change in the acceptance of Open practices as a legitimate way of teaching, not just a way to save students money.

We also delivered workshops both face to face and online in which participants were able to actively search for appropriate Open Textbooks. We delivered content about open licensing and its advantages for curriculum redesign and on how the process of adoption, adaptation and creation works. The workshops were open to faculty and instructors not just at the institution hosting us, but also others in the region who wanted to participate. This was done purposefully to ensure more system cross-collaboration, particularly within disciplines. It also had the benefit of making our thin resources spread further.

Finally, we presented to student groups. Getting students involved has been key, particularly at research institutions. In the early stages of the project we had interest from students, but not much activity. The concept was not only new to faculty, but to students as well. While the idea of saving money was obviously attractive to them, the intricacies of open licenses and other aspects of open practices were not yet familiar enough for students to truly begin advocacy work. As the project progressed, we were able to begin addressing this problem.

Student activity related to OER has now changed radically. Three years in, we now have a very active student movement in BC. Students are ensuring government knows how much they value the project by connecting with them directly. They are also doing their own advocacy work, including a twitter hashtag campaign #textbookbrokeBC which does a great job of highlighting the primary concern of students in the OER movement, the cost.

Learn from people who are doing what you're doing

We have also formed very strong relationships with individuals and projects in other jurisdictions who provided us with guidance in the early stages of our project. Connie Broughton, formerly with the Washington State Board for Community and Technical Colleges came to BC to tell us about their project and help us think through ours. Una Daly of the Open Education Consortium came to help us develop and deliver our first Open Textbook Workshop. Una and Connie as well as Paul Stacey and Cable Green of Creative Commons, Daniel Williamson and David Harris of OpenStax CNX, Dave Ernst of the Open Textbook Library, Nicole Allen and Nick Shockey of SPARC, Una Daly of the Open Education Consortium, Tricia Donovan of eCampus Alberta, Megan Beckett of Siyavula and David Wiley of Lumen Learning were all participants in the first Open Textbook Summit hosted by BCcampus in April 2013. This event, which began with a small group of passionate advocates with a will to collaborate has grown into an annual international event of nearly 200 participants in which faculty and instructors share their experiences and learn new ways of using open textbooks to advance learning. Once again we see that the strength of a collaboration, this time over international boundaries rather than just within a single province, is a vital part of this community and its ability to have an impact. There is a willingness among those in the Open community to work with each other which is unlike any other community I've been part of. There is a lack of ego and a desire to move forward which drives this culture of sharing. It is this fundamental paradigm that I think makes the movement unique in academia, and it is this uniqueness that gives me hope that the movement will continue to grow.

Adopt and adapt

We realized early on that it would be foolish to start creating books from scratch when there were so many others who had started before us on whose work we could build. We went through our list of 40 subjects and started looking at what was in the Commons that would meet our needs. Because of the work in other jurisdictions, like that of OpenStax CNX, College Open Textbooks and the Open Textbook Library, we were able to adopt many texts into our collection. We have continued to use this method of building out our collection as new titles are released by our partners. We realize many of the books will require adaptation to be fully usable by BC educators, but to give faculty a number of resources to choose from and potentially adapt seemed to us a better value proposition than simply finding one book for every subject area. This is particularly true given the different learning outcomes used in courses, even those that are taught on the same topics at the same institution. This reuse of resources has enabled us to stretch our project dollars to ensure that we had funding to pay when creations were necessary due to a lack of existing OER.

In the future we plan to do more of this type of resource use, but we will be more focussed on including ancillary resources to ensure faculty are able to build out full packages of the instructional resources they need using OER. When we have consulted with faculty, one of the barriers to adoption that they identify is that resources they use from traditional publishers often come with additional components such as PowerPoint presentations, teaching notes and exam banks. This is an area of our project in which Psychology faculty have lead. In Summer 2014, 17 Psychology faculty gathered together over a period of 2 days to produce a bank of nearly a thousand exam questions using a sprint model of development. These questions were then formatted by the BCcampus team in a way that makes them easy to reuse by faculty. This collegial event not only produced an excellent resource, it built community within the discipline around openness.

Bring an academic focus

We wanted to ensure an academic focus to our project so we began work early on to enable that. Having seen the review process being used by the Open Textbook Library, we adapted a set of review criteria used by the Saylor Foundation and implemented technology to support the process. Faculty were paid a small honorarium to do the reviews, and we post their feedback online for others to see and use in their evaluation of the textbooks. Posting the reviews enables other educators to understand how a given textbook might meet the needs in their course. It also has the added side benefit of bringing the reviewers into our community. By using reviews as a relatively non-committal way to allow faculty to dip their toes in the Open Textbook well, they become much closer to being prospective adopters, adapters, creators and advocates.

To further enhance the academic focus of our project we also implemented a Faculty Fellows Program. We put out a call for applications and ultimately selected 3 individuals who could represent the teaching perspective and their disciplines. Our Fellows were from 3 different institutions and represented the disciplines of Psychology (Dr. Rajiv Jhangiani), Philosophy (Dr. Christina Hendricks), and Chemistry (Dr. Jesse Key). The role of these individuals is to provide advocacy, research and advice to the project. They do this both within their own institution with faculty colleagues, senior administrators, students and others who have a role to play in OER practice as well as at discipline specific conferences, other institutions and organizations. The ability to have faculty speaking to their colleagues rather than our project team having those conversations is extremely powerful. The example of someone with relevant, current classroom experience who is working in the Open can be the impetus an instructor needs to start exploring that for themselves. Our first 3 Faculty Fellows exceeded our expectations in terms of the impact they had on the project, and as you can see from the reports each did following the program, it seems they had a similarly positive experience. Specifically, much of the work focussed on advocacy via presentations both internal to their institutions as well as with other organizations. The research component they undertook has also provided us with an invaluable view into actual practice. Dr. Jhangiani, our Psychology Fellow and co-editor on this book, contributed to a research project undertaken in a partnership between the OER Research Hub's Dr. Rebecca Pitt, BCcampus and our Faculty Fellows. This research documents the attitudes and practices of British Columbia Faculty with respect to OER. In addition, Dr. Jhangiani conducted his own research with students in his Psychology classroom. Hearing directly from students about their experience and attitudes toward OER is extremely useful in developing strategies for implementations.

What happens in the background when we're out there talking to people

As our OER colleagues will attest, the topic of quality has often dominated the discussion of Open Textbooks. When we have done presentations, the ways in which traditionally published textbooks are produced are often pointed to as the gold standard. In order to change this perception, we engaged a professional editing firm to work on our textbooks. While this was an extremely cost intensive aspect of our project, we felt that in order to counter these concerns and make the case that Open Textbooks can be of equal or better quality than those which are traditionally published.

We also expended a lot of resources cleaning up resources we had adapted (a process our staff fondly refer to as 'cleaning up crap'). We have combed through books replacing images that were not openly licensed, cleaning up tables and other formatting, adding tables of contents and fixing incorrectly done attributions. This work is not glamorous but it is extremely necessary if students are to get the maximum value out of the resources and if the resources are to be shared further. Releasing the books in this way ensures a strong foundation for any adaptations that will be done in future.

In addition to 'cleaning up crap,' we of course have to ensure a strong technical infrastructure. At BCcampus we have had a repository of resources with a web front-end for more than a decade. Most of those resources unfortunately get little use. There are many reasons for this, but one of them was most certainly the user experience, which was difficult to navigate and search. Because of this experience, we knew it would be important for the look and feel of our collection to be inviting and easy to navigate. Early on in our project we began to curate our collection by subject area and presented it via a clean, easy to use interface.

We implemented an ecosystem of technologies to support our project. This included Lime Survey for collecting review responses, Wordpress for the front end of our collection and Pressbooks, which is our authoring platform.

We recognized that we would need a platform on which we could adapt and create textbooks. We settled on an open source version of Pressbooks, a WordPress plugin. We have since done modifications to the functionality of that tool to make the work of faculty easier. Pressbooks also enables us to provide html versions of our textbooks in a format that has the look and feel of a textbook so that students can use the platform directly.

High touch support

Several of our Open Textbook team have experience as instructional designers and course developers, which explains our desire to not only replace traditional resources with Open ones, but our equally passionate stance on using Open practices to enhance learning. This background lead us to understand that a supportive approach would be key to helping the faculty we worked with make the shift to Open. During creation and adaptation projects, we provided project management, instructional design and technical support to faculty. We have two staff who are responsible for the project management role on our adaptation and creation projects. From the time we engage with an instructor, the project manager becomes their point of contact with BCcampus and is the person who guides them through the process. This support ranges from arranging contracts, developing specific deliverables and milestones, technical support, instruction on open licensing and often moral support. Faculty engaging in this process are often new to Openness. They are sometimes anxious about how their work will be received by colleagues. They are also often unfamiliar with the technical platforms being used and lacking critical knowledge about open licenses and publication. BCcampus staff approach these issues in a variety of ways, depending on the need. Sometimes it's a matter of connecting instructors with each other so they feel less alone, sometimes it is long phone calls to slowly step through technical training or the way in which open licences are assigned and how they work.

Through building these relationships, we have been able to help faculty through the process so that they don't feel so alone. We have also worked to ensure that institutions are aware of the work we are doing with their faculty so they may also provide support. We are now looking to build capacity in BC institutions so that teaching and learning centre staff, librarians and others are able to provide those supports that are so critical to enabling faculty to truly achieve the benefit of Openness. Several institutions in BC have recently formed OER working groups to develop their own pathways to supporting this work. At BCcampus, we are actively working toward ensuring instructional support staff in institutions are able to respond when faculty show a desire to work more openly. We want those staff to be able to answer questions about everything from licensing to what technologies to use (and how to use them). We also want them to be able to introduce the notion of not just OER, but Open Pedagogy and its value. We see Instructional Designers as crucial to this process given their role in the course development process and plan on working with them more closely in future.

We think it's working

We use a variety of factors to measure the success of our project. Some are very quantitative, for example how many adoptions, how many students impacted, amount of students savings.

Like others, we have struggled to find an accurate measure to inform our student savings numbers because of the complexities around rental programs and students who normally would not purchase a textbook. At this time, we use US$100 as our replacement value to average out what the actual costs might be along with the other factors influencing the reality of student purchasing activities.

Others success factors are more difficult to quantify but are measures we nonetheless value. We see our work as having value in terms of changing culture. Fundamentally I see Open Practices as a social justice movement. Institutional culture that truly values student access and learning is what we're shooting for. This means not only embracing OER, but also using the concept of Openness to its fullest potential. By taking advantage of the affordances of open licenses we have the ability to produce more relevant, engaging, contextualized learning experiences for students and that's where we should all be heading.

When we first started the project, we were operating in a world in which OER were largely unfamiliar to faculty in British Columbia. In the case of those who were aware, many were skeptical of the quality of resources, and indeed of the project itself. Three years later, much of that cynicism had disappeared. We rarely encounter hostile audiences these days, and when we do have someone in the audience who is dubious, it is often another audience member who provides a response before we do. As mentioned, many BC institutions are now forming OER working groups. These subtle shifts indicate success to us, and are in many ways more important than the numbers as they indicate an overall will in the system to support and value this work.

Do we need Policy?

Policy remains an interesting issue for us at BCcampus. Thus far, all of the work done by our organization has happened in the complete absence of formal policy. We have grant stipulations, but nothing more. To date, this has not been a barrier to us accomplishing our goals. However, at this stage of our project, when we are in a space of needing to build capacity out to our institutions so that ownership of the resources produced is with the BC system rather than with BCcampus, the question of policy is becoming more relevant. One of the targets we have for the next phases of our project is to assist institutions in developing their own policies that spell out how faculty will be supported in their OER use, the value of this work to the institution, and the expectations of quality in such an initiative. Some institutions are already beginning to work in this area, for example Simon Fraser University has made small-scale grants available to faculty who want to adopt an Open Textbook or other OER. At other institutions, OER working groups are being established. These examples are very positive, but we also need to move toward more formal mechanisms such as job descriptions that include work on open initiatives rather than having the work done off the side of people's desks.

Conclusion

As we move into the 3rd year of our project, we plan to continue to build capacity in our system and work toward more adoption of our resources and of Open practices in general. We have already begun to facilitate workshops on open pedagogy and curriculum redesign. It is my hope that this continued evolution of the conversation about open will draw more educators in and will ultimately result in Open as the default. I am very optimistic about the future of Open. The sheer volume of work being done is a clear indicator that if we continue to push as we have been, we will achieve our goals. That said, as with any movement, as we become bigger, we begin to fracture into groups with what may appear to be competing goals. An example of this is the conversation about whether the focus on Open Textbooks is doing damage to the movement overall, or whether Open Textbooks are indeed the door through which we will reach many more faculty who want to embrace Open. While I am completely in favour of being self-critical so that our growth isn't stagnated by patting ourselves on the back, I also think it's important to recognize and be respectful of the efforts of those who started the movement and those who are actively making change happen. What we need is action, collaboration, reflection and scholarly work that leads us to better outcomes overall.

TeachPsychScience.org: Sharing to Improve the Teaching of Research Methods

David B. Strohmetz[*], Natalie J. Ciarocco and
Gary W. Lewandowski, Jr.

Monmouth University

*dstrohme@monmouth.edu

Editors' Commentary

Designing a course to be engaging can be challenging, especially when the course in question is one that students often dread having to take and traditional sources of inspiration are limited, static, uneven in quality, or cost-prohibitive. In this chapter, authors David Strohmetz, Natalie Ciarocco, and Gary Lewandowski discuss the development of a website devoted to the sharing of openly-licensed and peer-reviewed strategies and demonstrations for teaching research methods and statistics. In doing so they demonstrate how individual faculty can marry the recognized resources and practices of their discipline, such as support from professional societies and peer review, with open licensing to be able to 'take a little piece of their world and improve it.'

Teaching research methods and statistics can be frustrating. The courses that are foundational to the undergraduate major, and psychological science in general, are the very ones that students often dread or avoid taking.[1] Even when they do take these classes, they do not necessarily see the value of this newly gained knowledge.[2] Hence, faculty teaching research methods or statistics in

How to cite this book chapter:
Strohmetz, D B, Ciarocco, N J, and Lewandowski, Jr. G W. 2017. TeachPsychScience. org: Sharing to Improve the Teaching of Research Methods. In: Jhangiani, R S and Biswas-Diener, R. (eds.) *Open: The Philosophy and Practices that are Revolutionizing Education and Science.* Pp. 237–244. London: Ubiquity Press. DOI: https://doi.org/10.5334/bbc.s. License: CC-BY 4.0

psychology often start every semester with a class full of less than enthusiastic students. This attitude does not seem to impede their learning. Most students achieve a baseline mastery of the material, but still leave the class with poor attitudes about statistics and research, and an inability to see their usefulness.[3] This puts faculty in a unique position. Perhaps more so than in any other course, faculty need support in improving students' learning experiences in these classes. In this chapter, we will describe our initiatives to support engaging research methods teaching, one of which is TeachPsychScience.org, an open-access resource to support teaching research methods and statistics.

The Challenge

When each of us first began teaching research methods and statistics, we experienced the same problem that many new instructors face. How should I structure my classes to make them both engaging and educational? How can I best promote student enthusiasm for the course? Initially, we did what we imagine other instructors do. We spent time reviewing textbooks, adopting ones that we thought students would find engaging and illuminating. We then constructed our lectures based on the content covered in each chapter, making sure that students memorized the terminology and mastered key concepts. As we gained experience from teaching these courses multiple times, we started to experience a restlessness in our approach. Importantly, the course was not broken or going terribly. In fact, the course was going well. Yes, students were learning. Yes, students gave us favorable course evaluations. However, they were not displaying the same passion for psychological science as we have.

After numerous conversations with our colleagues and soul searching, we began to realize that part of the problem was our approach to teaching. While our students were learning methodological and statistical concepts, they really were not engaging in the process of psychological science. We knew the value of active learning and wanted to include more in our classes, but we had a problem. High-quality teaching demonstrations, activities, and other similar resources to support our teaching of research methods and statistics were extremely limited. There are resources available, but each can be problematic for various reasons.

Instructor Manuals

One possible place to find these things are the instructor's manuals that accompany methods or statistics textbooks. There are several limitations to relying on instructor manuals. First, the purpose of these manuals is to facilitate the integration of the textbook into one's course, not necessarily improve the overall quality of learning and engagement in the scientific process in that course.

No text is perfectly comprehensive so there were often topics we include in our classes which are not necessarily covered in the text. Second, the quality of the activities and other materials in instructor manuals is uneven, limiting their utility. If the instructor materials simply reiterate the book's content, it is hard to truly broaden students' exposure to the material and mastery of the skills. Part of the issue may be that the instructor's resource authors tend to be different from the textbook authors and therefore may not be as experienced in teaching the course. The third problem is that even if a textbook has a quality instructor's manual, it is available only to adopters of that textbook.

Teaching Conferences

Attending national and regional teaching conferences is another way to garner resources. Each of us have gleaned creative teaching demonstrations and strategies from these opportunities. However, few of these conferences focus specifically on teaching methods and statistics. In addition, our limited travel funds meant we had to be judicious in our choices of which conferences to attend.

Teaching Journals

A third resource are teaching-oriented journals such as the *Teaching of Psychology* and *Scholarship of Teaching and Learning in Psychology*. These peer-reviewed journals publish a wide range of articles concerning teaching in psychology. Periodically, a compilation of articles specific to teaching research methods and statistics are published. A recent example is the e-book, *Teaching Statistics and Research Methods: Tips from ToP,*[4] freely available through the Society for the Teaching of Psychology (http://teachpsych.org/ebooks/index.php). These resources are invaluable, but searching opportunities for specific types of activities are limited and they are static in their content.

A Little Piece Project

In our department, we have an informal motto, 'Take a little piece of your world and improve it.' As we thought about our methods courses, we decided to do more than simply improve the quality of our individual classes. We wanted to take this 'little piece' of our world and improve the quality of instruction in research methods and statistics as widely as possible. We discussed ways for strategies for doing this and the type of support that was available for such open-access initiatives.

We started with a small project, capitalizing on the funding and dissemination opportunities provided by the *Society for the Teaching of Psychology* (STP). STP has an 'Office of Teaching Resources in Psychology' (OTRP),[5] dedicated to promoting the development of resources for advising and teaching psychology. To facilitate and financially support the development of teaching resources, STP annually offers up to five Instructional Resources Awards. We received one of these awards in 2009 to create a resource for instructors which identifies student-friendly research articles for illustrating various research design elements. We provided instructors with discussion starters and in-class activities to accompany each article. Our goal was to promote the reading of published research studies as students learned about the research process.

We were happy with the result and pleased that we had a means for sharing this open access resource with other instructors via the STP website.[6] However, we realized we could do more. Given the static nature of a pdf document, our resource was limited in its ability to be efficiently search for specific demonstration ideas. Likewise, we could not continually update this resource as we found new articles employing interesting or creative research designs. Besides, reading actual research studies is just one way in which students can engage in the process of science.

As our conversations continued, we realized that more often than not, when we needed a teaching idea, we would turn to one of our colleagues. We wondered if we could create a centralized repository where colleagues could easily share their ideas, suggestions, and demonstrations specific to teaching research methods and statistics. This website would be dynamic, allowing the posting of new ideas and teaching activities. It would also be easily searchable and provide opportunities for visitor feedback.

Online repositories to support instruction is not a new idea. MERLOT[7] is a very broad and general repository of resources available to instructors across a wide range of disciplines. As each of us teach social psychology, we can access more than 5,000+ links to teaching ideas and resources on Jon Mueller's 'Resources for the Teaching of Social Psychology' website.[8] The Social Psychology Network also has a page dedicated to Social Psychology Teaching Resources.[9] We wanted to create a similar website to specifically support the teaching of research methods and statistics. This open-access website would be textbook agnostic and continually updated with new contributions, minimizing the limitations of instructor manuals. It will provide an ongoing opportunity for instructors to share the ideas, strategies, and activities they use to facilitate student learning in these courses. The idea of TeachPsychScience.org was born.

TeachPsychScience.org

Creating and maintaining a website is a major undertaking. While one of us had experience with developing a website to support the dissemination of psychological research on relationships to the general public,[10] we were starting from scratch in the development of this website. We had several guiding principles for our endeavor. First, the resources available on the website had to be pedagogically sound. This meant that all submitted material would undergo a peer-review process before being posted on the website. Second, the website had to be user-friendly and searchable, making it easy for instructors to identify resources to meet their immediate teaching needs. Third, the website had to be both sustainable and scalable as we wanted this to become a valuable and ever evolving resource for all methods and statistics instructors.

When we considered issues surrounding copyright concerns, we decide to make all of the resources on the website subject to Creative Commons Licensing. We were not interested in monetizing the website; rather we simply wanted to help others become better teachers of statistics and research methods. We believe that instructors should have a right to modify or adapt the resources to meet their particularly teaching situation. Creative Commons Licensing provides this flexibility while still protecting the author from copyright infringement if the resource is used for something other than noncommercial educational purposes.

As we developed our vision for the website, we knew that finding financial support for this venture would be a major challenge. We are fortunate to have professional organizations in psychology which support pedagogical initiatives. We obtained a grant from the *Association for Psychological Science (APS) Fund for Teaching and Public Understanding of Psychological Science*. This grant funded the development and maintenance of the website for several years. Second, we needed to find a company to both build and host the site. We had a web designer[11] work with us to translate our ideas and vision into a viable website. Our goal was to provide a user-friendly website that allowed visitors to find resources in multiple ways, including searching by keywords or category (e.g., experimental designs or factorial analysis of variance). This meant that we had to spend much time planning the organizational structure of the website so that it would be both logical and intuitive to visitors. On the back end, we wanted an administrator interface that we could easy navigate in order to manage and update the content of the website.

Once we created the underlying structure for the website, we needed a core set of resources before we could actually launch TeachPsychScience.org. Drawing upon what we collectively do in our own classes, we created over 50 resources spanning the various categories of resource materials. These resources included class demonstrations, student practice exercises, writing assignments, class/lab activities, and links to other web-based resources supporting the instruction of research methods and statistics. We did not want the site dominated by our own contributions, especially as we wanted to glean ideas from our talented colleagues. We contacted colleagues at other institutions, asking them to submit any activities they used in their methods or statistics classes for inclusion in website. Interestingly, many responded by lamenting that they really did not have any good activities but wished they did. This provided further evidence of the need for the open-access repository we were creating. To expand the resources on the website, we monitored teaching of psychology publications and requested submissions from authors. We promoted the website at teaching conferences and teaching workshops and encouraged presenters and attendees to submit any activities or demonstrations they had to the website.

We believe that a key to the success of TeachPsychScience.org is to only include resources that were of high quality. To achieve this, we established a submission and peer-review process to ensure that all materials posted on the website are pedagogically sound. Having added this traditional peer-review process also provided contributors the justification for including their submissions on their curriculum vitae. We contacted scholars in the field of teaching and learning to serve as initial reviewers for the website. We also offered the position to those who contributed several high-quality submissions. To aid the process for reviewers, we created a rating form to accompany reviews and established some basic instructions for reviewers. With the help of between 5–8 reviewers and a number of additional submissions, we launched the website on June 1, 2010.

Open-access resources are valuable only if they are known and used. Our first challenge was spreading the word of the site's existence. We contacted other teaching resources sites, asking them to provide a link to TeachPsychScience. org. We advertised the website on various disciplinary websites and at teaching conferences. The site was featured in APA's *Psychology Teacher Network* Summer 2010 newsletter and is still promoted on APS's website. Over the past four and half years, the site has had over 48,000 users, 66% of whom were new visitors. Not surprisingly, traffic tends to be higher at the start of the fall and spring semesters. It is clear that this online repository for open-access resources is meeting a need among research methods and statistics instructors.

The second challenge we face is ongoing. The long-term success and value of TeachPsychScience.org depends on the continual submissions of resources by other instructors. Currently, there are over 100 demonstrations and other teaching ideas available for instructors to incorporate into their classes. The continued utility of this open-access resource depends the instructors sharing their creative ideas for promoting student learning in research methods and statistics. Currently, we rotate as editors for the website every 6 to 12 months. One of us takes the lead as the liaison between submissions and reviewers, makes final decision about the quality of a resource, and manages the site content.

Finally, at some point we will need to identify a stable funding source for the continuation of TeachPsychScience.org. Our initial grant provided financial support for the first five years of the website. We recently secured another grant from APS to support the expansion of the website to include resources to support the teaching of scientific writing. As writing is one of the three pillars of the research process, along with statistics and methodology, we felt this missing aspect was critical to develop and share. The additional funding provides support for material solicitation and development, along with a redesign of the website. The website update will allow for submissions type we did not originally consider (e.g., submissions that necessitate multiple accompanying files), incorporate writing-specific activities, and make the site more mobile friendly. This additional grant funding will also ensure the availability of the TeachPsychScience.org in supporting quality teaching for the next several years. Eventually, we will need to address the long term viability of the website.

Your Piece of the World

TeachPsychScience.org started with a simple desire – our wish to improve how research methods and statistics are taught so that all students may come to love the process of psychological science as much as we do. We are fortunate that the financial support we received first from the *Society for the Teaching of Psychology* and then the *Association of Psychological Science*, coupled with the Creative Commons License, brought our desire to fruition. It is through the open sharing of our collective wisdom and experiences that we were able to pick our little piece of the world and improve it as we continue to enhance and strengthen learning as students embrace psychological science.

What piece of the world are you going to improve? The fact of the matter is that anyone can replicate our efforts with TeachPsychScience.org in other areas of psychology. Think of the course you teach, your research expertise, as well as the psychology topics that capture your attention. In each of those areas you likely have something valuable to share with your fellow teachers and the field. If you have resources for teaching methods and statistics, we hope that you will share them with us. But more importantly, we hope that you find ways to extend your influence beyond your own classroom, into the classrooms of your colleagues and ultimately to improving students' lives.

Notes

[1] Rajecki et al., 2005
[2] Sizemore & Lewandowski, 2009.
[3] Sizemore & Lewandowski, 2009, 2011.
[4] Jackson & Griggs, 2012.
[5] OTRP, n.d.
[6] STP, n.d.
[7] MERLOT, n.d.
[8] Teaching of Social Psychology, n.d.
[9] Social Psychology Teaching Resources, n.d.
[10] Science of Relationships, n.d.
[11] Invisual, n.d.

References

Invisual. (n.d.). Available at http://www.invisual.us/

Jackson, S. L., & Griggs, R. A. (2012). *Teaching statistics and research methods: Tips from ToP*. Retrieved from the Society for the Teaching of Psychology Web site: http://teachpsych.org/ebooks/stats2012/index.php

MERLOT. (n.d.). Available at http://www.merlot.org

OTRP. (n.d.). Available at http://teachpsych.org/otrp/index.php

Rajecki, D., Appleby, D., Williams, C., Johnson, K., & Jeschke M. (2005). Statistics can wait: Career plans, activity, and course preferences of American psychology undergraduates. *Psychology Learning & Teaching, 4,* 83–89. DOI: https://doi.org/10.2304:plat.2004.4.2.83

Science of Relationships. (n.d.). Available at http://www.scienceofrelationships.com/

Sizemore, O. J., & Lewandowski, G. W., Jr. (2009). Learning may not equal liking: How a course in research and statistics changes knowledge but not attitudes. *Teaching of Psychology, 36,* 90–95. DOI: https://doi.org/10.1080/00986280902739727

Sizemore, O. J., & Lewandowski, G. W., Jr. (2011). Lesson learned: Using clinical examples for teaching research methods. *Psychology Learning & Teaching, 10,* 25–31. DOI: https://doi.org/10.2304/plat.2011.10.1.25

Social Psychology Teaching Resources. (n.d.). Available at http://www.socialpsychology.org/teaching.htm

STP. (n.d.). Available at http://teachpsych.org/resources/Documents/otrp/resources/ciarocco10.pdf

Teaching of Social Psychology. (n.d.). Available at http://jfmueller.faculty.noctrl.edu/crow/

DIY Open Pedagogy: Freely Sharing Teaching Resources in Psychology

Jessica Hartnett

Gannon University, hartnett004@gannon.edu

Editors' Commentary

*The philosophy of Open hinges on the simple act of freely and fearlessly shar-
ing one's creative works in service of the greater good. This intent is recognizable
even when not accompanied with the formal adoption of an open license. In this
chapter, author Jessica Hartnett describes the origin story of her 'not awful and
boring' blog, dedicated to the teaching of statistics and research methods. In doing
so she discusses some of the limitations of traditional outlets for sharing pedagogical innovations, provides practical tips for those looking to follow her lead, and
comments on the career implications of this approach.*

Many of the contributions to this book have involved large scale sharing of
open resources, the likes of which still involve large organizations (Open Science Framework, Noba, etc.). But there are simple, easier ways to contribute to
the free sharing teaching ideas without publishing your own open textbook. In
this chapter, I'm going to share with you my experience with two relatively easy
ways to share resources about the teaching of psychology. Mostly, I'm going to
talk about my blog,[1] dedicated to the teaching of statistics and research methods, but I will also talk some about my experiences as an editor at the Teaching
of Psychology Idea Exchange.

How to cite this book chapter:
Hartnett, J. 2017. DIY Open Pedagogy: Freely Sharing Teaching Resources in
    Psychology. In: Jhangiani, R S and Biswas-Diener, R. (eds.) *Open: The Philosophy
    and Practices that are Revolutionizing Education and Science*. Pp. 245–254. London:
    Ubiquity Press. DOI: https://doi.org/10.5334/bbc.t. License: CC-BY 4.0

Origin Story

I didn't set out to be a blogger about the teaching of statistics. Upon graduating with my PhD in 2009, I started teaching many sections of Psychological Statistics (essentially, Introduction to Statistics but from the perspective of someone with training in psychology) at Gannon University, a small liberal arts university in Erie, PA. I started with two sections per semester and now teach three statistics classes a semester, including Honors and Online versions of the class. Becoming a better statistics instructor is an important professional goal. One way that I wanted to improve upon the class was by incorporating modern, real life examples to reinforce statistical concepts. This approach worked for me across a variety of other psychology classes (Social, Developmental, Introduction) and while Psychological Statistics is quantitatively different (har-har) than other psychology courses, I still wanted to incorporate this teaching technique. Finding examples was easier than I expected: Once I had this goal to find relatable statistics examples, there were everywhere I looked. I'm a news junkie and a Facebook junkie, and statistics and data are enjoying a zeitgeist. For instance, look at the attention and criticism in October 2015 following the World Health Organization's reason classification of processed meats and red meats as carcinogens.[2] While on the surface, this is purely a global health story, but with a little digging, it becomes an example of relative versus absolute risk (What is one's likelihood of developing different kinds of cancer? Do most people consume a serving of processed or red meat per day?), operationalizing a variable (Do Spam and bacon really belong is the same category?), and whether or not proper research methodology was used (WHO made a casual claim based on largely observational research).

One can easily find stories like the bacon story by following certain entities on Facebook and Twitter, like Nate Silver, io9, Pew Research, etc. and paying extra attention to news websites for stories incorporating data. So, I managed to accumulate a bunch of fun examples but no evidence that they were enhancing student learning. As I was on the tenure track, I wanted to provide 1) evidence of teaching efficacy as well as 2) get a scholarship of teaching publication or two under my belt.

My very first foray into the scholarship of teaching and learning was a manuscript published in the journal *Teaching of Psychology* entitled 'Stats on the Cheap'.[3] It contained four original, empirically supported lessons for the teaching of statistics using free or inexpensive internet resources including http://www.gapminder.org/ and https://docs.google.com/forms. However, these four ideas represented just a few of the internet based resources I had found for teaching of statistics. Looking for statistics teaching resources had become a chronically accessible goal for me, and I soon found that I had resources and news stories and examples from a wide variety of largely pop sources, including the Huffington Post, Last Week Tonight, The New York Times, The Daily Show,

The Colbert Report, A.V. Club, Bloomberg Business, National Geographic, The Economist, and Mother Jones, just to name a few.

And my list of resources was quickly outgrowing the Google Document that I was using to organize them. I needed a better way to archive all of my materials. Additionally, the publication of Stats on the Cheap convinced me that there is interest in my approach to the teaching of statistics and I wanted to better serve my profession by sharing my teaching ideas more widely. I am a helper by nature, and I intimately understand the challenge of teaching statistics. Whether your statistics classes are aimed at just psychology majors or a broader student population, you are actively combating math anxiety when you get in front of your classroom. And you aren't just teaching statistics, you are teaching a framework for approaching scientific inquiry, lessons that are desperately needed in a world where statistics and research and polls are increasingly used as forms of rhetoric and persuasion. What we are doing is hard and important. That doesn't mean it can't be engaging and not awful.

So, in February 2012, I established my blog 'Not Awful and Boring,' featuring (to date) 167 free ideas/resources/news stories/web sites/etc. Note: While I don't teach research methods, I think it is impossible for a psychologist to divorce the research methods from statistics, hence, I advertise the website as serving research methods instructors as well. As of November 2015, the blog has been accessed over 70,000 times by users in 192 different countries.

As my blog has grown, I've come to believe that free sharing of teaching ideas via the internet may be the only way to keep up to date with the sharing of teaching resources. I think that because of a) I'm not convinced that every single teaching example needs to be rigorously assessed, b) the sheer volume of statistics resources available online, and c) the limitation of paper textbooks.

Publications are not always necessary for the sharing of 'small' teaching ideas

I believe that the scholarship of teaching is not as well respected as the more traditional scholarship of discovery. I think it is just as valuable and I think it is necessary. Before anyone implements large-scale changes in the way we think about pedagogy, student motivations, and ideal teaching methods, teachers need good science to backup best practices. For a strong example of why such efficacy research is very important, I suggest reviewing the 'learning styles' theory, one that has been disproven but is still widely taught.[4] More good examples for the necessity of teaching efficacy research can be found in the book 'Make it Stick,'[5] which is an in-depth review of the most effective ways in which people learn (interleaving, retrieval practice) and a rebuke against widely used methods (re-reading, highlighting texts) that are not as effective.

In this spirit, when I write scholarship of teaching manuscripts, I tend to concentrate on the assessment of larger, more formal lesson plans that I have used successfully across multiple semesters. Creating the manuscripts is a large time commitment, every bit as rigorous as the more traditional scholarship of discovery. If anything, scholarship of teaching can often time take longer, especially if an entire school year has to lapse in order to collect data from a control class as well as an experimental class.

However, I don't believe that I need to perform efficacy tests in order to demonstrate that each of my 187 blog postings aids in student learning. There aren't enough hours in the day to do so. But practicality is a lazy argument. Instead, I would argue that (1) overarching course assessments provide support for the small examples I provide and (2) teaching ideas are already being shared in public platforms without overly rigorous peer review.

Most universities are under increased pressure from accrediting organizations to formalize assessment and outcome measurement for individual courses. As such, I think that assessment of an individual meme or news story is included as part of the assessment of a class' efficacy. If learning outcomes for an individual instructor's class are met, and that class happens to include a few examples from my blog, then I think that speaks in a small way to the usefulness of the materials I share on my blog. Do I deserve congratulations for the hours of prep and grading and teaching that went into that class's success? Absolutely not. But did I maybe help a bit? I think so.

Another argument against the necessity for formal peer review for every possible teaching idea lays in the fact that there are already avenues for sharing teaching ideas that do not require extensive peer review and revisions of presentations.

When I present my pedagogy research at conferences, it is evaluated by a conference chair or a group of my peers, but not in the same, exhaustive manner that my publications are judged. I have never been asked to make revisions to any of my conference publications. Heck, I've given invited talks at prominent conferences that received absolutely no review.

Another reason that I would argue against the necessity of peer review is because it could also mean that useful, timely learning examples would be withheld from instructors and students. This is especially an issue with the teaching of statistics, because stats are hot right now.

Statistics is enjoying a zeitgeist, and I suffer from an embarrassment of riches.

Statistics is hot right now. Data storage is cheap, many of our social and financial transactions take place in ways that are easily quantified (shopping online, social media website, text messages, phone calls) leading to much data mining via 'big data.' More and more organizations are seeing the potential in data-driven decision making, which in turn leads to more and more popular sources writing about data and statistics at level that is accessible to an undergraduate. As statistics gets an increasingly extensive treatment from accessible sources, it is less and less feasible to vet every example. Most of the examples I provide address very specific issues related to statistics, but typically within a context that invites further debate about real life issues related to our use of data. For example, a recent posting from fivethirtyeight.com did a fantastic job of illustrating the balance between Type I and Type II error using not abstract ideas, but instead using parole decisions made by the Pennsylvania Department of Corrections. If you make your standards for parole too high, you are going to continue to imprison folks that would not have reoffended (false positive). If you make your standards for parole too lenient, you are going to release repeat offenders (false negative). So, this example illustrates a fundamental concern for all research scientists, but makes it more pertinent to our students by using a concrete example that can spur further discussion about social issues.

Additionally, even without popular sources, there is a lot of news happening in statistics. Use of p-values to determine statistical significance is increasingly under fire, up to and including being rejected by several large research journals. The replication crisis is another source of a lot of writing in statistics and research methods.

Text books take time to write and publish. Also, copyright issues.

Another way to share instructional resources would be via an edited text. However, many of the resources I suggest are proprietary and perhaps cannot be shared via text book for legal reasons. Additionally, given how long it takes to create a text book, many of the current events that demonstrate an application of statistics will not be current events by the time a book is published. Which leads to a loss of potential teaching aids.

So, those are my arguments for the free sharing of teaching resources. Now, I will provide some practical advice on how to actually do so. My blog has been a success, because I make it easy on my readers and on myself.

Sustaining a teaching blog

Give the people what they want

Presumably, you would like a readership for your blog. Easier said than done. It is difficult to make an impression in the blogosphere. There are already plenty of bloggers out there who share psychology content. Many of them have thousands of Twitter followers, some of them are rock stars in our field. Trying to add more to that noise may not be the best way for you to spend your time if you are motivated by the possibility of blogging to gain some free advertising/notoriety.

How have I sustained my blog and increased my audience? I provide a specific resource to college and high school teachers of statistics. My audience is largely overworked and they appreciate a one-stop repository that they can return to when they want to add content to a course or lecture. Not only do I have many examples to share, but they are organized by topic, making it easier for my readers to quickly find what they are looking for.

My regular readers also count on me as a consistent resource. I update my blog every single Monday, and sometimes more frequently. I also share additional links/retweets via the Twitter account associated with my blog.

In order to stay up to date with the blog, I give users the options to follow the blog via email, Google+, LinkedIn, and I maintain a Twitter account that is used solely for the purpose of announcing new blog posts and sharing tweets related to the teaching of statistics and research methods.

Make it easy for yourself

To make blogging about teaching easy, I need to return briefly to one of the ways of making your blog easy on your reader that also makes life easier for a blogger: Do not try to compete against pre-existing resources. Instead, figure out what unique, helpful contribution YOU have to make to serve your peers. I spend too much time on the internet. I enjoy helping and connecting with others. I love following the news. I teach a bunch of statistics classes. Finding material for my blog is not a chore. Frankly, it just means that I'm awake and probably need to spend less time on Facebook.

Because I know who I am and what I am good at, I have a pretty narrow focus and narrow audience. I'm not trying to be the biggest psychology blogger out there. I'm not wearing myself out by expanding my scope in order to increase my number of pages views. Perhaps this is limiting in some ways, but it keeps my work focused and allows me to serve my colleagues better. Again, I have been an overwhelmed statistics instructor, struggling to make my class relevant in the eyes of my students. As such, I have a sense of what my readers will like and what they will ignore (a sense that is informed by user data that Blogger collects). I also deeply believe that statistics are a very important topic for everyone to learn. If you are considering a blog, I would strongly suggest the advice from above: Figure out who you are and what you passionately desire to share with your colleagues. What unique skill or knowledge do you have to pass along? While I didn't do this, it may also be useful to see what other competition there is for your specialty area. For example, you may be very enthused about the teaching of social psychology, but can you realistically compete/create an audience/offer something unique and different from Jonathon Mueller's CROW website?[6]

So, don't reinvent the wheel. Similarly, use any and all technological shortcuts that you can. While my blog has become a way for me to share ideas with my peers, the origins of this blog are not entirely altruistic. I tried and failed to list all of my teaching resources is a big old Google Document. That became unwieldy and I decided to move all of my ideas into a blog, using the blog host Blogger.[7]

The decision to go with Blogger was based purely on the fact that I was already a Google user and Blogger is affiliated with Google and I was not motivated to do an exhaustive search for the very best blogging host on the internet but I am motivated to have one less password to memorize.

Another reason I went with a blog is because I did not want to create a website. I didn't want to learn HTML, or how to use an HTML editor, or work within the confines of a free website with the limited resources/advertising heavy. I didn't want to overly concern myself with aesthetics. I did not want to pay a web designer. So a free blog was a more reasonable format for me. Additionally, as I update this resource weekly, and every new resource is shared as a new blog post.

Another nice feature of Blogger is the ability to easily create and edit drafts of my blog posts. Let's say that I see an interesting Tweet from FiveThirtyEight. I copy the URL into a new blog post and I can save the post without publishing it. There are typically another two or three drafts and edits before the resource is ready for sharing. Sometimes, I use screen grabs from the original resource (when applicable) to illustrate why I think this is a useful resource to share. Then I queue up the blog post. Typically, I have about ten blog posts that are scheduled and ready/nearly ready for publication and another five or six that still require a fair amount of editing. This is especially nice when I know I'm going to be busy in at specific times in the future. For instance, I'm currently pregnant with my second child and I have several blog posts ready for publication right around his due date and for several weeks following his arrival.

Another motivation for blogging is the fact that I am comfortable with shameless self-promotion. I've been able to 'advertise' my own scholarship of teaching and learning publications, conference presentations, and a co-authored book. Rest assured, upon publication, this book will also be shared via the blog.

Alternative to a dedicated blog

There is a good chance that my recommendations for blogging your teaching resources actually convinced you to never, ever want to start a blog. Good on you for knowing yourself. However, this doesn't mean that you don't have resources to share.

There are other ways to provide service to your profession. I do so as a contributing editor to the Society for the Teaching of Psychology's Teaching of Psychology Idea Exchange.[8] My main contributions there are in Industrial-Organizational Psychology, a class I only teach every other year (and, thus, does not constitute a chronically accessible goal). As with statistics, many of these ideas come from news stories related to course content, such as a recent and highly publicized decision made by a large consulting firm (Accenture) to do away with annual evaluations.[9]

While I concentrate on I/O contributions, ToPIX includes contributions for every specialty area within psychology and each area has several categories (In the News, Audio/Visual, Classroom Lesson Ideas).

Anyone can apply to submit resources (email the editor at TOPIX@TeachPsych. org for instructions on how to do so). If you apply to submit resources, and do so frequently enough, you may be invited to become an Assistant Editor. This distinction is especially important for those of us on the tenure track, as there are many, positive ways that open sharing can demonstrate either scholarship or service to one's profession.

Career implications for open access teaching resources.

In addition to organizing my thoughts, sharing with my peers, and self-promotion, my blog and work with ToPIX serve my broader career goals. My employer recognizes and rewards the scholarship of teaching and learning. My blog meets my university's requirements that teaching scholarship must be professional, communicated, and peer-reviewed. Now, we know what these requirements look like when it comes to a traditionally edited and published manuscript (a presentation at a juried professional conference or a peer-reviewed article in a journal).

My blog is certainly communicated to the public. I have the Google Analytics data to demonstrate my growing readership. The content of my blog make it professional. No pictures of my dog or my fancy dinner whilst on vacation, just statistics. Additionally, if you peruse my Twitter following, you can see that the majority of my followers are scholars of data science and psychology. Finally, I have elicited peer evaluations for my blog. As such, I 'count' my blog as scholarship of teaching. However, you should proceed carefully. There are probably still people on your promotion and tenure committee that will have a negative knee-jerk reaction to you listing a blog on your application for advancement if you don't explicitly explain how your blog is a professional contribution.

My work as an Assistant Editor for ToPIX and the Society for the Teaching of Psychology is recognized as national service to my profession. Additionally, I have served on multiple committees for the Society for the Teaching of Psychology, which creates a cohesive narrative of the service I have provided for my profession.

Alternately, a teaching blog could also be considered national service to the profession. However you feel comfortable presenting you work for the purpose of promotion and tenure, I suggest you quantify your argument. I have Blogger as well as Google Analytic data about my blog. In addition to simple page views, this data demonstrates that my readership has increased over time as well as the fact that my blog has a global audience. I also have Twitter data that can demonstrate number of followers.

Conclusion

Teaching is hard. Teaching statistics is really hard, but made easier with engaging examples. We live in a time where many such examples are available freely and easily accessible via the internet. I have focused my efforts on sharing such examples that are relevant to the teaching of statistics, but there are plenty of other areas of psychology that could use a focused blog.

Notes

[1] Not Awful and Boring, n.d.
[2] WHO, n.d.
[3] Hartnett, 2012.
[4] Willingham, Hughes & Dobolyi, 2015.
[5] Brown, Roediger & McDaniel, 2014.
[6] Mueller, n.d.
[7] Blogger, n.d.
[8] ToPix, n.d.
[9] Cunningham, 2015.

References

Blogger. (n.d.). *Homepage.* Retrieved from https://www.blogger.com/home

Brown, P. C., Roediger, H. L., & McDaniel, M. A. (2014). *Make it stick: the science of successful learning.* Cambridge, MA: Belknap Press.

Cunningham, L. (2015, July 21). *In big move, Accenture will get rid of annual performance reviews and rankings.* Retrieved from https://www.washingtonpost.com/news/on-leadership/wp/2015/07/21/in-big-move-accenture-will-get-rid-of-annual-performance-reviews-and-rankings/

Hartnett, J. L. (2012). Stats on the cheap: Using free and inexpensive internet resources to enhance the teaching of statistics and research methods. *Teaching Of Psychology, 40*(1), 52–55. http://doi.org/10.1177/0098628312465865

Mueller, J. (n.d.) *Resources for the teaching of social psychology.* Retrieved from http://jfmueller.faculty.noctrl.edu/crow/

Not Awful and Boring. (n.d.). *Not awful and boring examples for teaching statistics and research methods.* Retrieved from http://notawfulandboring.blogspot.com/

Topix. (n.d.) Homepage. Retrieved from http://topix.teachpsych.org/w/page/19980993/frontpage

Willingham, D. T., Hughes, E. M., & Dobolyi, D. G. (2015). The scientific status of learning styles theories. *Teaching Of Psychology, 42*(3), 266–271. http://doi.org/10.1177/0098628315589505

World Health Organization. (n.d.). Q&A on the carcinogenicity of the consumption of red meat and processed meat. Retrieved from http://www.who.int/features/qa/cancer-red-meat/en/

Conclusion

You Can't Sell Free, and Other OER Problems

Robert Biswas-Diener

NOBA Project, robert@nobaproject.com

When I took the senior editor position at Noba—a service providing open materials for students and instructors of psychology (see Chapter 16, this volume)—I loved the idea of transforming education in general, and the field of psychology specifically. I assumed that a free, high quality textbook would be seen as more attractive than a costly counterpart and I worried that our small staff would not be able to keep up with the demand of instructors clamoring for our resources. Instead, I was greeted with alternating indifference and disdain. The indifference was largely the product of inertia. Instructors weren't looking to change course materials when their current materials were working just fine, at no cost to themselves. Understandably, switching books represented a huge amount of work for instructors including creating new lecture materials and tests. The disdain came from a deep suspicion of Noba and open materials in general. I shouted—metaphorically—until my voice gave out about how Noba materials didn't cost a cent. 'It sounds too good to be true,' I heard time and again. If I used the word 'free' in open posts to professional societies, in e-mails or at conferences, the reactions were even harsher. I was asked to quit spamming list serves and some groups were even reluctant to share news that Noba was handing out monetary grants to students. It was a frustrating time.

My team and I quickly realized that there were hurdles to overcome if we wanted to be effective in our educational mission. These hurdles were not necessarily related to the quality of our materials, to our brand reputation, or to

How to cite this book chapter:
Biswas-Diener, R. 2017. You Can't Sell Free, and Other OER Problems. In: Jhangiani, R S and Biswas-Diener, R. (eds.) *Open: The Philosophy and Practices that are Revolutionizing Education and Science.* Pp. 257–265. London: Ubiquity Press. DOI: https://doi.org/10.5334/bbc.u. License: CC-BY 4.0

our lack of professional connections within psychology. Instead, the greatest obstacles to sharing our resources were related to fundamental problems concerning open philosophy itself. In this chapter I would like to address three of these problems: 1) problems with the basic open narrative, 2) common misperceptions concerning open resources, and 3) problems concerning the best advocates for open. It should be noted that I do not think that I have arrived at the final answers or the most insightful conclusions regarding these issues. Instead, it is my hope that this chapter will plant the seeds of discussion and reflection that might subtly influence the ongoing battle for open.

Fixing the Narrative

In his book, *The Battle for Open*, Martin Weller (2014) presents a cogent argument that it is Silicon Valley—venture capitalists and technology companies—that have offered the most compelling contemporary narrative concerning education. In Weller's estimation this narrative can be summed up in the single, strong statement 'education is broken.' It is a powerful rallying cry because it is suggestive of so many verbs: disruption and revolution. Conspicuously absent from the discussion of both disruption and revolution is widespread reflection about the degree to which either actually will yield superior outcomes or tangible improvements. Even more damning, perhaps, is Weller's argument—one I find personally persuasive—that the 'education is broken' narrative is inaccurate. By way of counterpoint, consider the following:

- More people are being educated than at any time at history, and this includes secondary and tertiary education.
- There is greater gender equality in education than at any time in history.[1]
- Educators at all levels continue to advance their pedagogic methods and improve best practices.

Unfortunately, 'No, it's not!' falls short of being an effective response to the Silicon Valley story. This is, in part, because there are a number of legitimate problems with education. In the United States, for instance, there are a number of difficulties with funding public education and rising costs associated with higher education. In other countries, such as Australia, there has recently been a push toward increased testing, and many educators see this as a potential pedagogic mis-step.[2]

It would be wonderful if advocates of open education in general, and open practices in psychology education specifically, had an effective narrative of our own. For all its virtues 'open' has long been plagued with problems. The idea that products and services might be offered free of charge, for instance, is often associated with inferior quality or an outright scam. Also problematic are the very traditions of the academy itself. Many university instructors prize their own expertise above all else and, historically, there has been no greater vessel for this showcase of knowledge than in publication. To suggest that publications will not only be shared freely, but might actually be modified, is offensive to that tradition. Similarly, suggesting to many academics that their teaching methods and resources might be openly distributed and modified can be disquieting, suggesting as it does the potential obsolescence of some instructors.

I, myself, had a difficult time transitioning from a more conventional educational mindset to my current understanding of the benefits of open philosophy. For instance, when I first learned about Creative Commons licensure I reacted in a way that I have since learned is common: I was horrified. The idea that someone could re-write or re-sell my writing was anathema to my academic sensibilities. My attitude changed when I began to consider 'impact,' broadly defined, as an important outcome for my writing. Based on the traditional academic publishing model a few of my articles have made an impact: I have 13 publications with more than 100 citations each. It is hard to estimate exactly what that means but it is fair to say that a few of my ideas have been read by my peers and, perhaps, even influenced their thinking in some small way. Then again, a cover story I wrote for *Psychology Today* (2013) magazine was openly shared 18 thousand times on social media. An article I wrote on self-determination in the Montessori school pedagogy was published in an open journal and was downloaded more than 8,000 times. In one study conducted by the Research Information Network (2014) articles appearing in the journal Nature Communications were read twice as often as their non-open counterparts six months after publication. Similarly, the open publications received about one and a half times the number of citations than did their closed counterparts. If one measure of success is the amount a publication is actually read then removing barriers to reading it seems sensible to me.

To the extent that there are other psychologists who might be initially skeptical of open approaches—and they are legion—I argue here that open needs a compelling narrative of its own. Open, itself, is often treated like an adjective— as in, 'this open textbook is free for students'—rather than as a verb, as in 'if we open this course it will be available to people around the world.' In the first instance the word open is equated with being free as opposed to its more accurate meaning in which it includes greater potential for collaboration, innovation, and contextualization. So, what is it that open is doing for education? I believe that open is helping to spread education and here—in my opinion—is a potentially productive narrative: 'education is spreading.' Due to the intertwined factors of digital technology and open philosophy education is creeping out of its traditional repositories. The most progressive institutions are keeping pace by following suit: they are offering on-line and distance learning, adult learning and certificate programs, MOOCs, micro-credentials, and access to foreign campuses. Treating openness like an action feels dynamic and is a better banner than open as an adjective.

I am attracted to this particular narrative for several reasons. First, the notion that education is spreading is a far more positive message than education being broken. I do not mean to suggest here that we whitewash legitimate problems associated with education in general, or psychology in particular. Rather, I think the idea of the spread of education speaks to the unspoken mission of people attracted to psychology: we believe that our field is so fundamentally interesting, so worthwhile in its exploration of the human mind and in the creation of interventions, that we wish that it was available to everyone. While many psychologists feel overwhelmed at the prospect that their field may be broken they might be energized to learn that it is spreading.

Another reason the 'openness leads to spreading' narrative is so compelling is that it is suggestive of sharing. Keltner (2009) and others offer evidence that altruism is a fundamental aspect of human nature. By appealing to the better angels of this nature open education has a 'we're all in this together' leaning. In a field plagued by academic cat fights, big egos, critical peer reviews, and departmental politicking, the promise of open psychology offers a salve. It overtly suggests that we can all better achieve the most non-controversial missions of our discipline—high quality research, effective teaching, effective intervention—if we share our data, our methodology, our resources, and our experiences.

I would like to linger on the issue of sharing for a moment. In some people's minds the notion of sharing is synonymous with 'giving away.' In the same way that sharing a cup of sugar with a neighbor means that you have one fewer cups of sugar. The sharing implied by open educational philosophy is similar to the knowledge we share in the classroom. When we lecture students we do not find our own reserves of knowledge somehow depleted. At its best open practices allow people to retain control of their content or methods or data even as they allow other people to use them as well.

Finally, the spread of education may be appealing because of its suggestion—subtle or great—that this process is inevitable. Ask any editor at a major academic publisher and they will likely agree with this sentiment. They understand that the landscape of journals has changed radically in the last decade. These days, authors share PDFs of their work on private web sites. In fact, this common practice often occurs in violation of copyright law. Even so, many academics see such sharing as part of a broader mission to make an impact that outweighs the potential legal risks. These days, fee-based journals also compete against open journals and even against non-traditional outlets like blogs and news stories. There are scandals involving replication and p-hacking. In my experience, these editors have read the tea leaves and are scrambling to adjust their approach to the market. They are finding ways to ride these currents instead of fighting against them. The most successful universities, like the most successful publishing companies, will be those that accept the inevitable and find a way to work within the new open landscape.

Despite all this positivity it is likely that the education is spreading narrative will spike the blood pressure of many worried instructors. In an age where information is flowing freely and the collected publications of research psychologists are available on the internet many people worry about the role that academics will play in this new landscape. There are, I believe, three primary roles for psychologists in the future. In the information age there is more demand than ever for researchers to create new knowledge and act as 'upstream' agents in the knowledge pipeline. We are seeing early versions of this as increasing numbers of researchers blog about their work and that of their colleagues. In addition, some academics will serve as interventionists—therapists and consultants—helping to bridge concepts an applications. Finally, the nature of psychology instruction will shift. Academics will find less call to be the 'sage on the stage' offering content-packed lectures to large halls of note scribbling students. In their new incarnation, instructors will act as a 'guide on the side' in which they facilitate learning by leading discussions, activities, and otherwise curating knowledge. There will still be demand for teachers but teachers will not solely be 'tellers.'

Fixing the Common Misperceptions

In the few years I have spent as an advocate of open educational resources I have heard more than a few skeptical remarks. These remarks have come to me first-hand through e-mails and conversations at conferences. And they have come to me back channel as I have heard from members of professional bodies and department heads who confess that open education resources are a potential threat to individual and departmental revenue. Roughly, I divide skepticism into two categories: challenges to quality and challenges to control of content. Challenges to quality or common among people who are considering adopting materials such as open textbooks. Challenges to control of content are common to people who are considering creating open materials such as lectures or chapters. In both cases I would like to confess that I am sympathetic. I believe my colleagues are—fundamentally—no different than me. We all want the best for our students. We all want to reduce the stress of our respective workloads. We all want to earn an adequate income.

Quality

Many times I have heard skeptics of OER apply the adage 'you get what you pay for' with reference to free resources. There is an assumption that the processes that lead to high *priced* products are the same as those that lead to high *quality* products. It is a fallacy, however, to jump to the conclusion that free or inexpensively produced products, by contrast are of lesser quality. Many

OERs, including Noba, have sophisticated adaptive learning technologies and expert created instructor's manuals and other materials that hold up well in side-by-side comparisons. In fact, Noba materials are expensive to produce and our costs include editing, software, expert consultation and other expenses. 'Free to students' should not be equated with 'free to produce.' I believe that the concerns over quality are, in large part, a historical spillover from the early days of OER in which products may actually have been of lower quality. What needs to happen—and increasingly is—is that instructors considering using OERs need a formal rubric for evaluating their quality.

Control over content

In 2007 I was proud to publish my first book. It sold as well as niche academic titles do. Then, in 2009 I received an e-mail from a psychologist in Iran. He told me that he had taken the liberty of translating my work into Farsi and was releasing it in his country. Since Iranians are not subject to international copyright laws, he continued, he did not need to ask my permission. He did, however, want to know if I would be willing to write an introduction to the Iranian volume! It was the first time I experienced losing control of my own intellectual property. I worried about the quality of his translation. I worried about my own potential loss of revenue. I worried that he might somehow misrepresent my work and thereby impugn my reputation. Then, after consideration, I came to terms with those worries as I realized than any loss of income was almost certainly negligible and that this man's 'piracy' had likely expanded my impact by spreading my ideas to a corner of the world they might not otherwise reach.

Most people's concerns regarding losing control of their intellectual property or reputation are understandable in spirit but do not play out in fact. A large part of the openness in OER is related to removing obstacles to sharing information. If a researcher were to publish her paper on college student stress in an open format, for example, she would be able to share it widely. It would be accessible to students, lay people, reporters and colleagues who might not otherwise have subscriptions to a traditional journal. When I look at my own academic publication record, I have recently shifted to submitting more often to open journals. I am still attracted occasionally to the siren song of top tier journals and the prestige they bring but I have been personally swayed by the obvious impact of my open publications. This stickier concern is that someone might appropriate your work, impugn your reputation, or attenuate the quality of your writing. Unless you are a best-selling author or major public figure I would encourage you to take a deep breath on this point. The single most likely thing to happen if you were to openly license your work is that a well-intentioned colleague would add some research references or make cosmetic edits to enhance its readability.

Ultimately, I believe that the solution to the problem of OER's less than stellar reputation is going to be found in peer testimonials. There is increasing research attention being paid to these resources, and to the extent that the results of these studies act as an endorsement of OER, it will help bolster people's confidence in them. The more powerful route to persuasion, I argue, will be when a colleague down the hall gives OER their professional seal of approval. Whether it is a peer who has created open lectures or one who has adopted an open textbook the truth is their words hold more water than do mine. For this reason I believe that all of us who use OERs have an obligation to speak publicly about our experiences with them.

Who is the best advocate for OER?

It may seem strange to ask who is best positioned to advocate for OER. Given the academic benefits, the social justice consequences, and the economic solutions that OER represents it is tempting to suggest that everyone ought to speak out. Truth be told, there are better and worse sales people. Historically, people in the OER vanguard are a select group: we are not the 'establishment.' We are not the most famous clinicians or the most highly cited researchers. We tend not to be associated with the most prestigious universities in the world. In essence, we are radicals and experimenters. In this respect we bear some small resemblance to social activists at the beginning of many other movements. Like them, it can be easy to dismiss our collective racket as the chanting of a disgruntled few until we reach a critical mass.

I argue here that we will reach a critical tipping point—one in which OER shifts from being a fringe experiment to the being standard operating procedure—when the circle of advocates expands beyond the vanguard. When the advocate group includes students, famous psychologists, and administrators OER will have arrived. I mention students especially because they are, by definition, the ultimate reason for OERs. The lack of student participation in the discussions about the creation, adoption and use of OERs is perplexing. Largely, I think their absence reflects a tradition in which education is done to students rather than with them. I would love to see more students demanding open materials as well as having a hand in their creation. Similarly, I would love to see the most established psychologists—those with the highest citation counts and best reputations—endorse OERs. Although it might sound crass to compare our intellectual luminaries to the types of celebrity endorsements we see on television I will admit that I would be swayed in many academic matters by testimonials from my most admired colleagues.

And here—with the example of a changing publishing industry fresh in mind—we arrive at the final thorny question regarding OER advocacy: regardless of who endorses OER the decision to use them is sometimes out of the hands of individual instructors. In some institutions it is the department as a whole that makes such decisions. In addition, state and professional mandates may also be factors that affect the decision to adopt OERs. Among the uphill battles for OER is the fact that individual instructors who are willing to experiment with pedagogy might be hamstrung by departmental policy. Some departments have cleverly dealt with individual differences in their faculty by allowing individual instructors to make their own decisions regarding OERs such as open textbooks. Other departments have assigned a traditional textbook common to all courses but have simultaneously allowed an optional open text alternative. There is, unfortunately, no standard way forward. Instead, the issue of how open education is spreading will be sorted by each unique context in which it arises.

Conclusion

I would like to conclude with something other than a standard summary of the key points you have just read. You are intelligent and I trust that you understood this chapter well and that you will consider its points with the same seriousness that you reflect on all other points of your instruction. Instead, I would like to close on an exciting note. Open education is transforming old institutions in eye-popping ways. Because I work with Noba I will use the single example of textbooks. Traditionally, textbooks have been books. That is, they have been single bound hard-copy volumes that act as a survey of the most important information in a particular field. With the advent of open education textbooks were more likely to be digitized and more likely to be free of charge. An improvement to be certain but still a traditional view of a book.

Currently, we are able to experiment with new understandings of textbooks that allow for improved teaching. Noba, by way of specific example, allows every instructor to pick and choose the content she wants to include and arrange it in any order she chooses. She can make it available digitally or in hard copy. The advantages of digital technology also allow us to enhance the text with mouse-over technology that allows readers to see pop-ups of full references or key vocabulary terms. We now have the capacity for students to take adaptive learning quizzes inside each chapter as they read it. Most interestingly still, it is now possible for instructors to design a textbook with core content and for each individual student to customize this common core with supplemental chapters that are unique to his or her own interest. For the first time in history each student in class could have a truly individualized textbook written by experts but which they, themselves, customized.

This is just a single example of the ways that openness can lead to new developments in pedagogy. It is time that we quite asking about the quality of open education resources. Instead, we should be asking, 'Now that education is spreading, what are we going to do with it?'

Notes

[1] UNESCO Institute for Statistics, 2013.
[2] Bowden, 2014; Cashen et al., 2012.

References

Bowden, T. (2014). NAPLAN study finds testing causes students anxiety, program not achieving original goals. Retrieved from: http://www.abc.net.au/news/2014–05-19/naplan-study-finds-school-testing-program-not-achieving-goals/5463004

Cashen, J., Hornsby, D., Hyde, M., Latham, G., Semple, C., & Wilson, L. (2012). Say no to NAPLAN. Retrieved from: http://sydney.edu.au/education_social_work/news_events/resources/No_NAPLAN.pdf

Kashdan, T. B., & Biswas-Diener, R. (2013). What happy people do differently. *Psychology Today*, 46(4), 50–59, Retrieved from: https://www.psychologytoday.com/articles/201306/what-happy-people-do-differently

Keltner, D. (2009). Born to be good: The Science of a Meaningful Life. New York: W.W. Norton & Company.

Research Information Network. (2014). Nature communications: Citation analysis. Retrieved from: http://www.nature.com/press_releases/ncomms-report2014.pdf

UNESCO Institute for Statistics. (2013). Adult literacy rate, both sexes (% ages 15 and older). Retrieved from: http://hdr.undp.org/en/content/adult-literacy-rate-both-sexes-ages-15-and-older

Weller, M. (2014). Battle for Open: How openness won and why it doesn't feel like victory. London: Ubiquity Press. DOI: http://dx.doi.org/10.5334/bam

Open as Default: The Future of Education and Scholarship

Rajiv S. Jhangiani

Kwantlen Polytechnic University, rajiv.jhangiani@kpu.ca

The opposite of open is not closed; the opposite of open is broken. The more I think about it, the more this cogent observation, made by John Wilbanks, resonates with me

Scholarly publishing is certainly broken. Here, the farce that passes for tradition supplements public funding for researchers with generous dollops of (publicly subsidized) voluntary peer review and editorial work. The taxpayer is then asked to provide additional funding for database subscription fees so that institutions can access the very research they produce. And as if paying three times was not enough, if the very same taxpayer wished to access the fruits of all this labour, they would instead find a paywall. That is, unless the researcher had access to even more public funding to cover (often exorbitant) article processing charges (APCs).

Science is arguably broken. Here, tradition incentivizes trading off unsexy but cumulative research for flashy but non-reproducible findings. Worse still, the prevailing system encourages questionable research practices like p-hacking and withholding disconfirming data. Every new generation of scholars learns that prestige is associated with communicating in the least accessible style through the least accessible and impactful channels.

How to cite this book chapter:
Jhangiani, R S. 2017. Open as Default: The Future of Education and Scholarship. In: Jhangiani, R S and Biswas-Diener, R. (eds.) *Open: The Philosophy and Practices that are Revolutionizing Education and Science.* Pp. 267–279. London: Ubiquity Press. DOI: https://doi.org/10.5334/bbc.v. License: CC-BY 4.0

Pedagogy too reveals many chips and cracks. Here, faculty routinely adapt their courses to map onto the structure of textbooks instead of the other way around. Lecturing remains popular, despite masses of empirical evidence that unequivocally show the advantages of higher impact practices such as active and experiential learning. Instructors still regularly assign 'disposable' assignments in which students produce work for one person while they in turn take pains to provide thoughtful feedback that is almost immediately recycled at the end of the term. And a great many educators continue to teach in a manner that assumes their principal role is that of content delivery, despite living in an age of unparalleled access to information.

Higher education, itself – if not broken – is certainly delusional. For how else can we describe an enterprise in which we continue to pretend that our students start and finish at the same place and at the same pace? Where we cling to the fantasy that our students have unfettered access to required course materials. Where our programs do not serve the modal student, who works at least part-time and will no longer spend four years studying full-time at the same institution. And where we claim to value being 'student-centered' when in practice faculty, course content, accreditation or testing requirements, and budgetary concerns drive the learning process far more than students.

All of this is why I bristle when I hear the old 'if it ain't broke, why fix it?' argument. For if it's not open, it is broken, and that's precisely why we must fix it.

Open Access as Default

A number of changes are needed in order for open access to become default practice. Government and other organizations that fund scholarship need to lead by requiring grantees to adopt open licensing for their publications (see Chapter 3 by Cable Green for the example of the US Department of Labor). At the very least, granting organizations need to mandate that copies of published manuscripts be placed in an open repository, even if after a short embargo period (see for example recent policies from the US National Science Foundation Open Access Policy or the Canadian Tri-Council Open Access Policy). More education and awareness is also needed with the issue of open access. Many scholars do not understand the differences between green (immediate self-archiving), gold/hybrid (APCs), and platinum (no APCs) open access. And still more equate open access journals with predatory journals that are happy to accept (in exchange for APCs) an article consisting solely of the phrase 'Get me off your fucking mailing list' repeated for ten pages (with accompanying flow charts and scatter plot graphs[1]).

Scholars need to work determinedly and collectively to challenge the status quo, to found and manage high quality open access journals. Chapter by Aaron Jarden and Dan Weijers provides an excellent example of this, as does the recent case of the mass resignation of the editors and editorial board of

Lingua due to Elsevier's refusal to change its APC policies.[2] In this latter case, the very same editorial team has since founded *Glossa* (published by Ubiquity Press, also the publisher of this book), a high quality open access journal with significantly lower APCs. Of course, given that scholars produce the research, provide the peer review, make the editorial decisions, and pay for the APCs, you might rightly think that we hold all of the power. But breaking with tradition in this fashion still requires taking a principled stand. This is most powerful when senior scholars (who typically hold grants and have the ability to pay high APCs) speak on behalf of junior (e.g., pre-tenure) scholars, who face the maximum pressure to publish in the highest impact journals in their field, and which may not provide an open access option. The irony of course is that articles published with an open access license are far more likely to be read and cited (SPARC Europe, 2016), and therefore actually have an impact.

Open Science as Default

Brian Nosek's description (Chapter 7) of a conversation with his young daughter about the differences between how science ought to function and how it actually does is the clearest and most evocative account I have come across of just how broken science is. I believe that if open science is to become default practice it will take a more radical shift, one driven by an ideological commitment to scientific progress[3]. Pre-registering one's hypotheses and data analysis plans will serve as a guard against *p*-hacking and hypothesizing after the results are known (HARKing). Sharing raw data in an open repository will deter the fabrication of data and the selective deletion of outliers. Sharing research materials and statistical syntax openly will facilitate replication. And designing and publishing careful, iterative, high-powered studies instead of single study publications with low statistical power will enhance rigour and replicability. Each of these practices needs to be incentivized or otherwise encouraged by the leadership and professional bodies within a discipline. Joining the more than 50 organizations and 500 journals that have adopted the Transparency and Openness Promotion (TOP) guidelines would be a good start.

In my discipline the *Association for Psychological Science* has been among the groups leading the shift to open[4], with its flagship journal *Psychological Science* awarding digital badges to researchers who adopt such practices. Looking at individual scholars, however, there has been noticeably greater openness to open science among those at the Assistant or Associate Professor level. Interestingly, many of these young leaders, like Simine Vazire, Sanjay Srivastava, Daniel Lakens, and Will Gervais, are also active bloggers, reflecting on methodological issues and their resolution squarely in the public domain. And while it is easy for traditionalists to discount blogging as 'not real academic writing,' these posts are widely shared and read, generating what is arguably a rapid and open peer review via posted comments.[5] As with writing op eds, blogging is a high impact form of open scholarship (note that the average journal article is read completely by about 10 people;),[6] one that is disincentivized by the academy and not easy or comfortable.[5] It requires an ability to think deeply, write accessibly, and publish fearlessly, and is motivated precisely by a commitment to scientific progress.

That same commitment to scientific progress may also spur the mainstreaming of open peer review, which, despite potential drawbacks like greater difficulty locating reviewers and a slower peer review process, produces reviews that are of the same or higher quality while being more courteous.[6] Indeed, according to Kriegeskorte:

> Open evaluation goes hand in hand with a new culture of science. This culture will be more open, transparent, and community controlled than the current one. We will define ourselves as scientists not only by our primary research papers, but also by our signed reviews, and by the prior work we value through our public signed ratings. The current clear distinction between the two senses of 'review' (as an evaluation of a particular paper and as a summary and reflection upon a set of prior papers) will blur. Reviews will be the meta-publications that evaluate and integrate the literature and enable us as a community to form coherent views and overviews of exploding and increasingly specialized literatures. (2012: 12)

Open Educational Resources as Default

For open educational resources (OER) to become the default choice for the majority of faculty they will first need to learn of their existence. Unlike traditional publishers' textbooks, OER do not have a well-oiled marketing machine – there are no unsolicited and cumbersome packages clogging faculty mailboxes, no offers to sponsor research conferences or student events, no smiling faces knocking on office doors. Moreover, once faculty learn of the existence of OER, they must interact with and review these resources to combat the common myth that what is freely accessible must be of low quality (forgetting of course that they are not free to produce). Indeed, OER have come a long way from the days of long text-based .pdf documents with no images, learning aids, multimedia, or interactive features. OER are now increasingly commonly supported by a robust range of ancillary materials like test banks, lecture slides, and instructor manuals, which is why 85% of 2,366 students and 2,144 faculty (aggregated across 9 peer-reviewed studies) familiar with OER now rate their quality as the same as or better than traditional resources.[7]

Given the exorbitant cost of textbooks in North America, it is understandable that open textbooks have been sailing primarily under the flag of cost savings. However, all of the advantages of open textbooks need to be clearly understood and articulated, for although the cost to students is a factor that (some) faculty pay attention to, it is far from the primary consideration when adopting course materials. Indeed, one major survey suggests that it may be the least important criterion to adopting faculty.[8] So faculty need to understand that while their students benefit from free, immediate, portable, and permanent access to required course materials, they too stand to benefit from the additional permissions to revise, remix, update, and contextualize the textbook to serve their pedagogical goals, and even embed and scaffold course assignments within the readings. And far from harming students, *every single one* of the thirteen peer-reviewed studies that has investigated OER efficacy has found that students using these resources perform the same as *or better than* those who are assigned traditional textbooks (see: http://openedgroup. org/publications).

Looking beyond course performance, data from a number of studies even show the positive impact of OER adoption on the number of credits enrolled in during subsequent semesters, and improved student retention and program completion rates.[9] Taken together, OER thus represent a big win for students, a win for faculty, and even a win for institutions. This is why, despite the absence of traditional marketing, OpenStax books are now adopted by one out of every five degree granting institutions within the United States (see Chapter 17), open textbooks from the BC project are now adopted at all 25 public post-secondary institutions in the province, and institutional membership in the Open Textbook Network continues to grow rapidly. OER is slowly but surely going mainstream.

Open Pedagogy as Default

Driving an airplane down the road. That is the metaphor that David Wiley uses when describing traditional, 'disposable' assignments, ones that 'students complain about doing and faculty complain about grading. They're assignments that add no value to the world – after a student spends three hours creating it, a teacher spends 30 minutes grading it, and then the student throws it away'.[10] Of course this assumes that the student retrieves the assignment at all, and, if they do, that they even briefly consider the feedback that the instructor has thoughtfully crafted.

While I take David's point, I confess that I do see value in many kinds of traditional assignments. I also recognize that most faculty aim to create engaging and authentic learning experiences for their students, ones that will allow them to become both knowledgeable and skillful, and to become good global as well as good local citizens. There is no shortage of good intent here, which is precisely what makes me optimistic. Although the use of fact-based multiple-choice questions drawn directly from publisher-supplied question banks is ubiquitous, a growing number of instructors are looking to harness the energy, potential, and even the creativity of their students in order to have them produce resources for the commons. However, the challenge of this approach is in designing assignments that:

1. allow students to develop and exercise useful skills that align well with course and program learning outcomes;
2. produce something that will add value to the world;
3. produce something that will be openly available;
4. provide sufficient support so that the experience will not be terrifying for them (a serious concern as we ask them to step outside of their comfort zones); and
5. build in enough latitude so that the assignment constitutes a creative project and not simple a recipe for the same product.

As seen in Chapter 9 by Robin DeRosa and Scott Robinson, open pedagogy and especially the notion of students as creators inhabits a transformational and inspirational space in which faculty do not simply adopt open educational resources but instead adopt open educational *practices*. It is this kind of transformative thinking that lead Delmar Larsen and many of his colleagues at the University of California at Davis to harness the efforts of thousands of students over several years to first build ChemWiki (now the most visited Chemistry website in the world) and what is now the massive LibreTexts library. A source of customizable course materials and learning analytics for dozens of institutions, the LibreTexts library leaves behind the archaic practice of static 'editions' of textbooks, with these rapidly outdated snapshots of a discipline now replaced with a living, dynamic, flexible, and interactive body of knowledge. And although the resource is demonstrably efficacious[11] and the cost savings for students are substantial, these traditional arguments for OER are now the encore as the pedagogy takes centre stage.

If open pedagogy is ever going to go mainstream several questions remain: is OER adoption a gateway to open pedagogy, as some claim? Given that it is far easier to place technology in people's hands than it is to get people to do things differently, how many potential OER adopters would actually take advantage of the license to revise and remix, or even involve their students in OER creation? If some are attracted to OER principally out of a concern for social justice whereas others are drawn by the potential for pedagogical innovation, the overlap between these two sets may well be labeled a gateway. But surely people differ not only in their awareness of open practices but also in their openness to practice. What are the implications of this heterogeneity within the open community and among our audience? These are all strategic and empirical questions that the field needs to grapple with.

Colonisers and Edupunks Unite

Shortly after the 2015 Open Ed Conference in my hometown of Vancouver, BC, Robert Farrow (a co-author of Chapter 5) wrote a blog post reflecting on the emergence of 'Coloniser' and 'Edupunk' subcultures within the open education community, those who aim to displace traditional textbooks by emphasizing *free* versus those who aim to transform pedagogy by emphasizing *freedom*. The language of social justice, access to resources, and open textbook efficacy belong to the former whereas discussions about innovative teaching practices, access to ideas, and open pedagogy belong to the latter. Evolution in education versus revolution in education.

Robin DeRosa, undoubtedly among the revolutionaries and a co-author of Chapter 9, had implored us to 'stop fetishizing the textbook, which is at best a low-bar pedagogical tool for transmitting information. OER is better than that.' Amanda Coolidge (Senior Manager, Open Education at BCcampus), on the other hand, reminded us that although 'we come to open for different reasons . . . yet for students, it is about cost. We have to remember that although we may be at the stage in open where we need to start talking and implementing open pedagogy many in the movement still care deeply about reducing student costs, and those are our student leaders. Students care about access and for students access to education means reducing financial barriers.'

Of course as framed here this is certainly a false dichotomy, as most people will fall somewhere in between the extremes of the Coloniser-Edupunk spectrum, and the pragmatism of the Colonisers likely affords the idealism of the Edupunks. I agree that there is something unsettling about promoting a free and open version (even with the '5R' permissions to reuse, revise, remix, retain, and redistribute) of a resource that itself is a dinosaur and in desperate need of rehabilitation. As David Wiley notes in Chapter 15, 'Until the full stack of our intellectual infrastructure becomes open, truly democratized innovation and permissionless innovation will be impossible.' However, it is also true that open textbook adoption does allow students to gain in terms of both cost savings and educational outcomes (at least for those who would not have otherwise purchased an exorbitantly priced 'required' textbook). And as Amanda points out, these outcomes are incredibly valuable.

What is more, *Edupunks vs. Colonisers* is not the only example of subcultures within the open education community. Take the case of Creative Commons licenses, which, to the uninitiated, can resemble hieroglyphics (see Cable Green's chapter in this volume for a key). Working from within the movement, however, can make them appear more like gang signs, with purists (including some contributors to this volume) vehemently advocating against the adoption of the non-commercial (NC) clause for reasons that range from a lack of clarity about what it means to a respect of the right to sell or profit from OER (including as a pathway to sustainability for the movement). On the other hand, there are others (including many new initiates) who adopt more restrictive licenses because they derive comfort from knowing that their work cannot be stolen and sold by traditional publishers (NC) or modified by others, potentially sullying their reputation if errors are introduced (no-derivatives; ND). And while it may simply take greater experience in the open arena to enhance confidence and reduce territoriality (once again, an empirical question), I am convinced that the open education movement will benefit from welcoming as many into our big tent as possible. Doing so necessitates respecting the creator's choice of license without judgment or snobbishness and recalling that the movement itself is predicated on inclusiveness, freedom, and generosity.

In my opinion what would be far more useful than forming factions with different battle cries would be recognizing and responding in a nuanced fashion to the heterogeneity present in our *audience*. This is an analysis for which I have found the pencil metaphor (an adaptation of Rogers' theory of diffusion of innovation for the ed tech context) to be especially useful (see Figure 1).

The *leaders* adopt a new innovation, driven by intrinsic motivations and willing to experiment and fail. The *sharp ones* learn about what the leaders are up to, get excited by the proof of concept and begin to adopt the innovation themselves. Together the leaders and the sharp ones form small pockets of innovation that persevere despite the absence of support (and sometimes in the face of opposition).

As an advocate for openness, my audience is usually the *wood*—the ones who represent the mainstream (you grip the wood of the pencil, after all). These include scholars who 'would' publish openly if the highest impact journals of

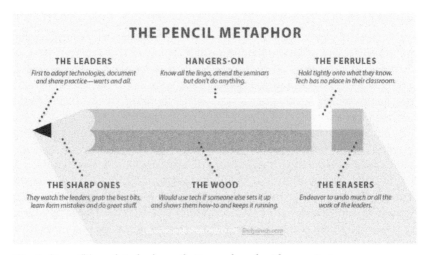

Fig. 1: 'Pencil' by sylviaduckworth. Reproduced with permission.

their field were open access or did not demand high article processing charges to grant open access. Those who 'would' share their raw data or research materials (but only when directly requested). And instructors who have never heard about OER but who 'would' adopt open textbooks if they were available for their subject area, were easily accessible, were accompanied with a suite of ancillary resources, were of sufficiently high quality, and had demonstrated efficacy. Or so they say. For one challenge with advocacy involves distinguishing the wood, who may have sincere, rational criteria that they wish to be satisfied prior to going open, from the *ferrules*, who raise fairly disingenuous objections (e.g., where is the efficacy research for OER?). You know, those criteria that, once met, are swiftly replaced by another set. Together with the *erasers*, the ferrules are occasionally the most vocal in the group.

One of the strangest things about the battle for open is that so much territory, so much of our teaching and learning environment, has long-since been ceded. This is why describing the ability to revise course materials to map onto one's pedagogical goals as a layer of academic freedom carries a feeling of novelty. Or why when an academic department's textbook selection committee shortlists half a dozen titles, allows for idiosyncratic blackballing, and ends up with the text everyone can all live with, this does not feel like a concession. Because for most faculty these are not spaces in which there is any real expectation of academic freedom. While this acutely reveals the intimate relationship between one's philosophy of teaching and one's openness to openness, herein lies the key to establishing open as default:

For faculty who enjoy experimenting and innovating with teaching, OER adoption is indeed a meagre position to advocate. These are the folks who will enjoy playing with authentic and open pedagogy, who may actually take full advantage of the ability to revise and remix, and understand that adopting open educational *practice* is really just about good pedagogy and in that sense is not at all radical. Similarly, for those writing and practicing at the cutting edge of science, facilitating greater sharing—of data, materials, and articles—is likely to be greeted with gratitude and even relief. Scrutinizing the wood, I observe faculty who currently adopt high-priced, static textbooks but care enough about their students to feel guilty about this decision (principled agents in a principal-agent dilemma?). In at least some of these cases, the ensuing guilt leads them to bend the course to map onto the textbook. While not an example of great pedagogy, this could be construed as an empathic response that ameliorates both their guilt and their students' resentment. This is the region of the wood where the social justice case for open textbooks may resonate particularly well. The same might be said for making a principled case for open access among scholars who already (illegally) share copies of their pay walled publications with their peers and on websites like Academia.edu or Researchgate.

There are numerous gateways to open, many ways in which this core philosophy and common set of values may be activated.[12] Adopting OER may serve as a gateway to adapting or even creating OER. Adopting OER for one course may serve as a gateway to adopting OER for other courses. One faculty member within a department adopting OER may serve as a gateway to other faculty adopters. One department adopting OER may serve as a gateway to other departmental adoptions within the same university or even region, or to the development of 'Z degree' programs. Adopting OER may be a gateway to open pedagogy. Open access publishing may serve as a gateway to OER adoption. And practicing open science may serve as a gateway to practicing open pedagogy. *Vice versa, in omnibus causis, ad infinitum.*

A Final Comment

I am unabashedly an optimist about the future of open. I believe that this shift is not just desirable and moral but also inevitable. After all, open access would be a good idea even without prohibitive APCs, open science would be a good idea even without data fabrication, and OER would be a good idea even without exorbitant textbook costs, although each of these problems makes the shift more urgent. At their heart, both education and science are about service through the creation, sharing, and application of knowledge, goals that are compromised by traditional practices that are closed and broken. This is why I believe that the future of both education and scholarship is open. I see it on the horizon.

Notes

[1] Mazieres & Kohler, 2005.
[2] Jaschik, 2015.
[3] We are already in the midst of a massive shift from the use of exorbitantly priced and proprietary (and surprisingly deficient) statistical software packages like IBM's SPSS to free and open source (and remarkably powerful) software packages like *R*; however, this shift appears to be primarily driven by its favourable position on the cost and quality axes.
[4] The APS has also supported the open access publication of this book, through a small grant from its Fund for Teaching and Public Understanding of Psychological Science.
[5] Heleta, 2016.
[6] van Rooyen et al., 1999; Walsh et al., 2000.
[7] Hilton, 2016.
[8] Allen & Seaman, 2014.
[9] Fischer et al., 2015; Hilton & Laman, 2012.
[10] Wiley, 2013.
[11] Allen et al., 2015.
[12] Weller, 2014.

References

Allen, E., & Seaman, J. (2014). *Opening the curriculum*. Babson Survey Research Group. Retrieved from http://www.onlinelearningsurvey.com/reports/openingthecurriculum2014.pdf

Allen, G., Guzman-Alvarez, A., Smith, A., Gamage, A., Molinaro, M., & Larsen, D. S. (2015). Evaluating the effectiveness of the open-access ChemWiki resource as a replacement for traditional general chemistry textbooks. *Chemistry Education Research and Practice, 16*(4), 939–948. Retrieved from http://pubs.rsc.org/is/content/articlehtml/2015/rp/c5rp00084j

Biswas, A. K., & Kirchher, J. (2015, April 11). Prof, no one is reading you. *The Straits Times*. Retrieved from http://www.straitstimes.com/opinion/prof-no-one-is-reading-you

Center for Open Science. (2015). *The transparency and openness promotion guidelines*. Retrieved from http://cos.io/top

Coolidge, A. (2015, November 23). Riffin' on the open textbook. [Web log comment]. Retrieved from https://gotcurls.wordpress.com/2015/11/23/riffin-on-the-open-textbook/

DeRosa, R. (2015, November 20). Open textbooks? UGH. [Web log comment]. Retrieved from http://robinderosa.net/uncategorized/open-textbooks-ugh/

Farrow, R. (2015, November 28). Colonisers and edupunks (&c.): Two cultures in OER? [Web log comment]. Retreived from https://philosopher1978.wordpress.com/2015/11/28/colonisers-and-edupunks-c-two-cultures-in-oer/

Fischer, L., Hilton, J., Robinson, T. J., & Wiley, D. A. (2015). A multi-institutional study of the impact of open textbook adoption on the learning outcomes of post-secondary students. *Journal of Computing in Higher Education, 27*(3), 159–172. Retrieved from http://link.springer.com/article/10.1007%2Fs12528–015-9101-x

Gervais, W. (2016). Homepage. [Web log comment]. Retrieved from http://willgervais.com/blog/

Government of Canada. (2015, February 27). *Tri-Agency open access policy on publications*. Retrieved from http://www.science.gc.ca/default.asp?lang=En&n=F6765465–1

Heleta, S. (2016, March 8). *Academics can change the world – if they stop talking only to their peers*. Retrieved from https://theconversation.com/academics-can-change-the-world-if-they-stop-talking-only-to-their-peers-55713

Hilton, J. (2016). Open educational resources and college textbook choices: A review of research on efficacy and perceptions. *Educational Technology Research and Development. 64*, 1–18. DOI: https://doi.org/10.1007/s11423–016-9434-9 Retrieved from http://link.springer.com/article/10.1007/s11423–016-9434-9

Hilton III, J., & Laman, C. (2012). One college's use of an open psychology textbook. *Open Learning: The Journal of Open, Distance and E-Learning, 27*(3), 265–

272. https://doi.org/10.1080/02680513.2012.716657 Retrieved from http:// scholarsarchive.byu.edu/cgi/viewcontent.cgi?article=1069&context= facpub

Jaschik, S. (2015, November 2). Language of protest. *Inside Higher Education.* Retrieved from https://www.insidehighered.com/news/2015/11/02/editors- and-editorial-board-quit-top-linguistics-journal-protest-subscription-fees

Kriegeskorte, N. (2012). Open evaluation: A vision for entirely transparent post-publication peer review and rating for science. *Frontiers in Compu- tational Neuroscience, 6.* DOI: https://doi.org/0.3389/fncom.2012.00079 Retrieved from: https://www.ncbi.nlm.nih.gov/pmc/articles/PMC3473231/

Lakens, D. (2016). The 20% statistician. [Web log comment]. Retrieved from http://daniellakens.blogspot.ca/

Mazieres, D., & Kohler, E. (2005). *Get me off your fucking mailing list.* Retrieved from http://www.scs.stanford.edu/~dm/home/papers/remove.pdf

National Science Foundation. (2015, March 18). *National Science Foundation announces plan for comprehensive public access to research results.* Retrieved from https://www.nsf.gov/news/news_summ.jsp?cntn_id=134478

Open Education Group. (n.d.). *Publications.* Retrieved from http://opened group.org/publications

Scholarly Open Access. (2016). *List of publishers.* Retrieved from: https://scholar lyoa.com/publishers/

SPARC Europe. (n.d.). *The open access citation advantage service.* Retrieved from http://sparceurope.org/oaca/

Srivastava, S. (2016, August 23). The hardest science. [Web log comment]. Retrieved from https://hardsci.wordpress.com/

Stafford, T., & Bell, V. (2012). Brain network: social media and the cognitive scientist. *Trends in Cognitive Sciences, 16*(10), 489–490. https://dx.doi.org/10.1016/j. tics.2012.08.001 Retrieved from http://www.tomstafford.staff.shef.ac.uk/ docs/stafford_and_bell2012.pdf

Thierer, A. (2014, November). Embracing a culture of permissionless innova- tion. Retrieved from http://www.cato.org/publications/cato-online-forum/ embracing-culture-permissionless-innovation

van Rooyen, S., Goodlee, F., Evans, S., Black, N., & Smith, R. (1999). Effect of open peer review on quality of reviews and on reviewers' recommendations: A randomised trial. *BMJ Clinical Research, 318,* 23–27. Retrieved from https:// www.researchgate.net/profile/Richard_Smith61/publication/13415327_ Effect_of_open_peer_review_on_quality_of_reviews_and_on_reviewers'_ recommendations_a_randomised_trial/links/54100e410cf2df04e75a55a6.pdf

Vazire. S. (n.d.). Sometimes I'm wrong. [Web log comment]. Retrieved from http://sometimesimwrong.typepad.com/wrong/

von Hippel, E. E.(2005), Democratizing Innovation. Cambridge, MA: MIT Press. Retrieved from http://web.mit.edu/evhippel/www/democ1.htm

Walsh, E., Rooney, M., Appleby, L., & Wilkinson, G. (2000). Open peer review: A randomised controlled trial. *The British Journal of Psychiatry, 176*(1),

47–51. DOI: https://doi.org/10.1192/bjp.176.1.47 Retrieved from: http://bjp.rcpsych.org/content/176/1/47.long

Weller, M. (2014). The battle for open: How openness won and why it doesn't feel like victory. London, UK: Ubiquity Press. DOI: https://dx.doi.org/10.5334/bam Retrieved from http://www.ubiquitypress.com/site/books/detail/11/battle-for-open/

Wikipedia. (2016). *Principal-agent problem.* Retrieved from https://en.wikipedia.org/wiki/Principal%E2%80%93agent_problem

Wilbanks, J. (2010, August 13). *The unreasonable effectiveness of open data.* Retrieved from https://dspace.library.colostate.edu/bitstream/handle/10217/81403/CI_Days_2010_Wilbanks.pdf

Wiley, D. A. (2013, October 21). What is open pedagogy? [Web log comment]. Retrieved from http://opencontent.org/blog/archives/2975

Index

A

Abelson, Hal 11
academic competencies 48
academic libraries 187, 191
    see also librarian views of open
        educational practices.
academic prestige, and open access
        publishing 188
academic publishing 182
    see also commercial textbooks.
    taxpayer paying four times
        for 185, 267
academic resistance to change
    see departmental resistance to
        change.
academic tribalism 169
access 5
accessibility 213
active learning 213, 214
adaptation 72, 74, 76, 231
    concerns about
        inappropriate 141, 262

adaptive learning 213, 214
adult education 48, 49
affiliation problems, and open access
        publishing 190
Africa 14, 18, 20
Allen, Fred 170
Allen, Nicole 231
American Psychological
        Association 50, 56, 144, 242
American Sociological
        Association 119
ancillary resources 83, 212, 222,
        232, 270
Angner, Erik 190
animations 127
Apache Software Foundation 104
Apple Keynote 126, 127
Aragon, Michelle 101, 110
article processing charges
        (APCs) 267, 269
Asia-Pacific Economic
        Cooperation 15

Aspy, D.   51
assessment   48, 71
  open   205
assignments   271
  information competency   145
  screencasts   135
  Wikipedia   118
Association for Psychological
      Science   96, 119, 241, 242,
      243, 269
'At School' (painting)   126
Atkins, Daniel E   15
attributions   144
author royalties   141, 151

B

badges on journal articles   93, 94,
      269
Baraniuk, Richard   10, 11
Battle for Open, The (Weller)   258
BC Open Textbook Project   142
BCcampus   17, 174, 228
  see also BC Open Textbook
      Project.
BCOER (BC Open Education
      Librarians)   229
BC Open Education Librarians
      (BCOER)   229
BC Open Textbook Project   34,
      227, 271
  academic focus   232
  adopt and adapt strategy   232
  ancillary resources   232
  consultation and outreach   229
  history of   228
  measuring success   234
  policy   235
  providing support   234
  quality   233
  relationships   230
  review process   232
Beasley-Murray, Jon   119
Beckett, Megan   231
benefits of OER   70, 115

financial   70, 71, 73, 75, 115, 212,
      224, 235, 270
  for educators   70, 73, 76
  for formal learners   72, 73, 74
  for informal learners   71, 72, 74
Berliner, D.   50
big publisher textbooks
  see commercial textbooks.
Blogger software   251, 253
blogs   246, 249, 252, 270
Bollier, David   31
Braun, A.   4
Brilliant Star framework   49
British Library   136
Brookhart, Sarah   96
Broughton, Connie   230, 231
Brown, John Seely   15
Bryant, Fred   190
Budapest Open Access Initiative   10
bulleted text   127

C

California Open Educational
      Resources Council   17
California State University   10
Campus Alberta OER initiative   34
Camtasia software   130
Cardiff Metropolitan
      University   191
Carnegie-Mellon University   11, 15
Cathedral and the Bazaar, The
      (Raymond)   108
CC0 tool   36
Center for Open and Sustainable
      Learning (COSL)   198
Center for Open Science (COS)   89
  cultural change   95
  evidence   93
  incentives   94
  infrastructure   95
  origins of   90
  training   94
Cerego   214
chalkboards   126

'Challenge of Defining Wellbeing,
    The' (Dodge)   191
Change   15
Character through the Arts
    project   54
Chemistry LibreTexts library
    173, 272
ChemWiki   272
China   15
China Open Resources for Education
    (CORE)   14
Chu, Lucifer   14
Cialdini, Robert   80
citations, traditional versus open
    access publishing   259
CK-12   17
Clarck, David   109
class cancellations, and
    screencasts   135
class-created textbooks   120
classroom technology   126
cognitive knowledge, focus on   49
Collaborative for Academic, Social,
    and Emotional Learning
    (CASEL)   49
collaborative learning activities   52
College Open Textbooks   231
Colonisers and Edupunks   273
colonization   4
Columbia University   10
commercial textbook authors
    resistance to OER   151
    royalties   141, 151
commercial textbooks   209
    ancillary resources   83, 212, 222
    author royalties   141, 151
    costs of   68, 149, 154, 209, 211
    criteria for choosing   170
    efficacy   81
    instructor manuals   238
    new editions   210
    non-traditional economic
        model   211
    price and quality   80

replacing use of   133, 200
    size   211
    student preferences   81
    versus OERs   82
committees, resistance to
    change   171
Common Core State Standards,
    US   17, 48
Commonwealth of Learning
    (COL)   15, 18
Community College
    Consortium   149
competency profiles   56
concept mapping   52
Connexions   10, 14
Coolidge, Amanda   273
cooperative independence   103
cooperative learning   52
copyright concerns   140, 187, 241,
    259, 262
copyrighted material, in
    screencasts   136
copyright law   30, 260
costs
    article processing charges
        (APCs)   267, 269
    commercial textbooks   68, 149,
        154, 209, 211
    gold open access   183, 268
    homework software access
        codes   156
    journal subscriptions   191
    minimizing user acquisition
        costs   224
    online only journals   182
    open textbooks   212, 214
    platinum open access   185, 191,
        268
    see also financial benefits of OER.
    taxpayer paying four times for
        academic publishing   185, 267
Cote, Jean Marc   126, 129
course assistants, screencasts
    for   135

Creative Commons   11, 12, 14, 17, 19, 231
Creative Commons Aotearoa New Zealand (CCANZ)   104
Creative Commons licenses   13, 30, 31, 197, 241, 274
    Attribution (CC-BY) license   12, 34, 38, 39, 68, 104, 105, 149, 222
    Attribution-No Derivatives (CC-BY-ND) license   13, 36, 37
    Attribution-Non Commercial (CC-BY-NC) license   13, 35
    Attribution-Non Commercial-No Derivatives (CC-BY-NC-ND) license   36, 37, 187
    Attribution-Non Commercial-Share Alike (CC-BY-NC-SA) license   170
    Attribution-Share Alike (CC-BY-SA) license   35, 105
    CC0 tool   36
    MOOCs   16
    Public Domain Mark   36
    Wikimedia Commons   136
Creative Commons search engines   136
creativity   6
credibility   5
critical digital pedagogy   117
critical reflection by educators   70, 76
crowd-sourced syllabus   121
cultural change   95, 163, 235
    and committees   171
    encouraging departmental control over OERs   170
    encouraging departmental culture of openness   167
    focusing on quality OERs   169
    understanding resistance to change   165

D

Daly, Una   231
decision-making errors   166, 249

Delissio, Lisa   54
Delmar Larsen   272
departmental resistance to change   163
    committees   171
    encouraging departmental control over OERs   170
    encouraging departmental culture of openness   167
    focusing on quality OERs   169
    see also resistance and indifference to OER.
    understanding   165
DeRosa, Robin   273
Diagnostic and Statistical Manual of Mental Disorders   210
Diener, Carol   170
Diener, E.   49
Diener, Ed   170
diffusion of innovation theory   172, 274
digital rights management (DRM)   223
digital textbooks   212
    see also open textbooks, advantages of.
direct instruction   50, 51, 52
discussion activities   52
distance education   58
distributed ecosystem model   223
document cameras   127
Dodge, Rachel   191
Donovan, Tricia   231
Dreyfus, Hubert   196
Dropbox   136

E

early drafting   142
East Georgia State University   215
eCampus Alberta   231
educational inequality   3
    see also equal access to education.
Educational Testing Service (ETS)   205

educators
  benefits of OER   70, 73, 76
  critical reflection   70, 76
Eich, Eric   96
Eldred, Eric   11
Elsevier   269
EngageNY   17
Enlightenment ideals   164
e-portfolios   50, 56
equal access to education   70, 110
Ernst, Dave   156, 231
evaluation of benefits of OER   67
  benefits claimed   70
  educators   73
  formal learners   72, 73, 74
  informal learners   71, 72, 74
exam proctors, screencasts for   135
experiential learning   53, 54
extra-curricular activities   44

F

Facebook   136
Facilitative Teaching   51
false positive and negative
        errors   166, 249
Farrow, Robert   273
financial benefits of OER   70, 71, 73,
        75, 115, 212, 224, 235, 270
first in family university
        learners   110, 155
Fischer, Lane   200, 204
flashcards   83
flexibility   5
Flickr   136
flipped classroom   51, 56, 119, 130,
        221, 222
focus of education   48
formal education   44
formal learners, benefits of OER   72,
        73, 74
Foundation for Learning Equality
        (FLE)   19, 20
free and open source software (FOSS)
  see open source software.

free cultural works approved open
        licensing   106
Freiberg, H. J.   51, 52
funding proposals, transparency
        in   106
future of education and
        scholarship   267
  open access as default   268
  open educational resources as
        default   270
  open pedagogy as default   271
  open science as default   269
  uniting factions and
        subcultures   273

G

Gabrieli, C.   49
Gage, N.   50
Gates Foundation   16, 17, 200, 201
Gervais, Will   269
Glossa   269
GNU-Linux operating system
        103, 108
Golan, Lawrence   31
gold open access   183, 268
Google Analytics   252, 253
Google Drive   136
Google Scholar   191
graduate student instructors,
        screencasts for   135
Green, Cable   231
green open access   268
Groom, Martha   119

H

Hamilton College, New York   190
Hammond, Allen   15
Harris, David   231
Harvard University   11, 191
Helliwell, John   190
Hendricks, Christina   232
Hewlett Foundation   9, 11, 12, 13,
        14, 16, 17, 18, 19, 68, 69, 70
Hewlett, Walter   11

Higher Education Authorization Act
    (2008), US   155
Hilton, John   204
homework software access codes   156
human rights   3
Hummel, J.   49
hybrid courses   130
Hypothes.is software   120

I

immigration   4
incentives for increasing
        openness   94
    badges on journal articles   93,
        94, 269
    Registered Reports   94, 96
    TOP Guidelines   94, 269
India   4, 15, 18, 20
indigenous learners   110
individualization of content   213,
        214, 264
informal education   44
informal learners, benefits of
        OER   71, 72, 74
information competency   145
infrastructure   14, 19, 95, 233
inquiry-based approaches   221
Institute for Open Leadership   39
Institute for the Study of Knowledge
        Management in Education
        (ISKME)   11, 14, 19, 20
instructor manuals   238
intellectual property concerns   140,
        187, 241, 259, 262
intellectual property policies   104
International Development Research
        Centre, Canada   19
International Journal of Wellbeing
        (IJW)   181
    and affiliation problems   190
    conception of   181
    Creative Commons licenses   187
    gold open access   183
    impacts of   190

open access publishing   183
open review policy   193
platinum open access   185, 191
prestige barrier   188
publication of source data   192
start up and funding   189
internet Archive   14
Ismail, S.   54
iTunes   136
iTunes U   136

J

Jefferson, Thomas   11
Jhangiani, Rajiv   167, 232
JISC (Joint Information Systems
        Committee), UK   68
Joseph, Heather   16
journal article badges   93, 94, 269

K

K-12 OER Collaborative   17
Kahneman, Daniel   190
Kaleidoscope project   201
Kelly, Kevin   149
Keltner, D.   260
Kerr, Clark   165
Key, Jesse   232
Khan Academy   16, 19, 35, 74,
        119, 136
Khan, Salman   119
Kozol, J.   4
Kraut, Alan   96
Kriegeskorte, N.   270

L

La Belle, T.   44
Lakens, Daniel   269
Larsen, Delmar   173
Laura and John Arnold
        Foundation   92
learned class   4
learner-directed pedagogy   117
    class-created textbooks   120
    crowd-sourced syllabus   121

learning tasks   52
Ledonne, D.   135
Lessig, Lawrence   11, 149
LibOER Listserv   149
librarian views of open educational
        practices   139, 147, 229
    early drafting   142
    giving credit   144
    information competency   145
    librarian roles   153
    putting students at center   145
    sharing   140
    studying licenses   144
    supportive feedback   143
    textbook costs   154
    understanding faculty resistance
        and indifference   150
LibreTexts library   173, 272
lifelong learning   50
Lime Survey software   233
Lingua   269
LinkedIn   56
localization of content   70, 213
Lumen Learning   17, 143, 201, 231

M

Marentette, Paula   119
Margulies, Anne   11
Massachusetts Institute of
        Technology (MIT)   10
    MIT OpenCourseWare (MIT
        OCW)   10, 13, 15, 18, 35,
        68, 197
Massive Open Online Courses
        (MOOCs)   16, 20, 69, 75, 148
Mellon Foundation   11
MERLOT   10, 220, 240
meta-cognitive strategies   52
Microsoft PowerPoint   126, 127
MIT OpenCourseWare (MIT OCW)
        10, 13, 15, 18, 35, 68, 197
MOOCs
    see Massive Open Online Courses
        (MOOCs).

Moodle-based OER   173
Mountain Heights Academy   199
Mozilla Foundation   104, 206
Mueller, Jon   240, 251
multimedia design   127
    see also screencasts.

N

NACSCORP   224
National Science Foundation
        10, 93
National Science Foundation Open
        Access Policy, US   268
Nature Communications   259
Ngugi, Catherine   14
Noba Project   82, 120, 142, 170, 174,
        213, 257, 264
non-formal education   44
norm-based assessments   49
Northwest Vista College   215

O

Oades, Lindsay   190
OER
    see open educational resources
        (OER).
OER Africa   14
OER Commons   220
OER Knowledge Cloud   69
OER Research Hub   67, 70, 232
OER universitas (OERu)   101, 104,
        106, 174
    lessons learned   111
    open design and
        development   108
    open pedagogy   110
online charter schools
        198, 199
online only journals   182
    *see also International Journal of
        Wellbeing* (IJW), costs.
Online Program Development Fund
        (OPDF)   229
online search engines   191

OOPS organizations   14
open access publishing   10, 16, 19,
    69, 183
  and affiliation problems   190
  becoming default practice   268
  citations   259
  gold open access   183, 268
  green open access   268
  open review policy   193
  platinum open access   185, 189,
    191, 268
  prestige barrier   188
  publication of source data   192
  see also International Journal of
    Wellbeing (IJW).
Open Assessments   205
Open Badges Infrastructure   206
Open Broadcaster Software   130
Open Competencies   205
Open Content License   197
Open Credentials   206
Open Culture   19
open design and development   108
Open Education Commons   136
Open Education Conference   15
Open Education Consortium   231
Open Education Global
    Conference   15
open education licensing
    policies   37
Open Education Research
    Group   76
Open Education Resource
    Foundation (OERF)
  see also OER universitas (OERu).
Open Education Resource
    Foundation (OERF)   102
open educational practices   102, 139
  becoming default practice   271
  early drafting   142
  giving credit   144
  information competency   145
  librarian roles   153
  putting students at center   145

  sharing   140
  studying licenses   144
  supportive feedback   143
  see also librarian views of open
    educational practices; open
    pedagogy.
open educational resources (OER)
  see also benefits of OER; open
    textbooks; resistance and
    indifference to OER.
open educational resources (OER)
  addressing key deficiencies of   221
  ancillary resources   222, 232, 270
  becoming default choice   270
  best advocates for   263
  common misperceptions   261
  control over content   141, 262
  defined   12, 196
  discoverability   222
  hardcopy versions   83
  history of   9, 14
  numbers of publications   69
  problems with narrative   258
  quality and price   262
  textbook metaphor   205
  types of publications   69
  versus big publisher books   82
open file formats   110
open governance   105
Open Government   19
Open High School of Utah
    (OHSU)   199
Open Journal Systems   183
open licensing   29, 102, 116, 197
  free cultural works approved   106
  importance for educators   73, 76
  importance for learners   72
  studying   144
  see also Creative Commons
    licenses; open education
    licensing policies
open pedagogy   110, 115
  becoming default practice   271
  class-created textbooks   120

crowd-sourced syllabus 121
lessons and challenges 121
Noba Project Student Video
Awards 120
privacy concerns 121
Wikipedia assignments 118
open philanthropy 102, 105
open planning 105
open policy 102
Open Policy Network 39
Open Policy Registry 39
Open Polytechnic of
New Zealand 190
Open Publication License 197
open review policy 193
open science 269
see also Center for Open Science
(COS), becoming default
practice.
Open Science Framework 92,
93, 95
Open Society Foundation 16
open source software 102, 103, 106,
110, 112
journal publishing 183
open source software
foundations 104
Open Textbook Library 231, 232
Open Textbook Network 156, 271
Open Textbook Summit (2013) 231
open textbooks 10, 17, 18, 68,
149–50 270
accessibility 213
active and adaptive learning
213, 214
addressing key deficiencies of 221
advantages of 212
ancillary resources 222, 232, 270
and digital rights management
(DRM) 223
and resistance to change 172
costs 212, 214
discoverability 222
efficacy 172, 199, 271

financial benefits 212, 235, 270
individualization of content 213,
214, 264
learner-created 120
minimizing user acquisition
costs 224
Noba Model 213
piloting use of 157
versus commercial textbooks 82
see also BC Open Textbook Project;
OpenStax.
Open University UK (OU UK) 11,
14, 19, 148
OER Research Hub 67, 70, 232
OpenLearn 68
OpenCourseWare (OCW)
see also MIT OpenCourseWare
(MIT OCW).
OpenCourseWare (OCW) 198
openness in education and
schooling 45
open pedagogy
becoming default practice 271
OpenShot Video software 130
open source software 12, 35, 196
OpenStax 10, 17, 68, 142, 149, 174,
219, 231, 271
addressing key deficiencies of
OER 221
ancillary resources 222
and digital rights management
(DRM) 223
Creative Commons licenses 222
discoverability 222
distributed ecosystem model 223
minimizing user acquisition
costs 224
mission support fees 225
quality 221
standard scope and sequence
requirements 221
OpenVA conferences 149, 150
Orange Grove 220
organisational closedness 103

Organization of Economic
    Cooperation and
    Development (OECD)   12,
    14, 15, 17, 19, 39
organizational structure   55
Otago Polytechnic, New
    Zealand   103, 112
overhead transparency
    projectors   127

P

Pakistan   20
parental expectations   51
peer review
    arguments against necessity
        of   247
    open review policy   193
    Registered Reports   94
    TeachPsychScience.org   242
Peer to Peer University (P2PU)   20
pencil metaphor   274
Perens, Bruce   197
personalized learning   70
*Perspectives on Psychological
    Science*   96
Petrides, Lisa   20
p-hacking   267, 269
PhET science simulations   18
Pierce College   143
Pitt, Rebecca   232
platinum open access   185, 189,
    191, 268
PLOS   10
podcasts   129, 133
policy
    impacts of OER on   71, 75
    open   102
positive psychology   49, 58
poverty, and educational
    inequality   3
Powdthavee, Nattavudh   189, 190
Preregistration Challenge   95
preregistration of analysis   95, 269
presentation software   126, 127

    see also screencasts.
pre-service training for teachers   20
PressBooks software   120, 233
prestige barrier, open access
    publishing   188
privacy concerns   121
problem-based learning   51, 56
Project Implicit website   91
project-based learning   51, 56
Psychological Science   93, 96, 269
Psychology Today   259
PsycWiki   173
Public Domain Mark   36
Public Knowledge Project
    (PKP)   183
purpose of education   47
    quality   79, 221, 233
    aesthetics versus effectiveness   200
    and efficacy   81, 199
    and price   80, 199, 200, 262
    focusing on   169
    OERs versus big publisher
        books   82
    perceptions of   151, 199
    student preferences   81

Q

QuickTime software   131

R

Raymond, Eric   108, 197
reciprocal teaching   52
RecordIt software   130
Reeves, Tom   204
Registered Reports   94, 96
reproducibility of scientific
    research   92, 93, 94
Research Information Network   259
research methods teaching, sharing
    ideas   237, 245
    arguments for free sharing   247
    blogs   246, 249, 252
    career implications   252
    challenges   238

instructor manuals   238
teaching conferences   239
Teaching of Psychology Idea
        Exchange (ToPIX)   252, 253
teaching-oriented journals   239
TeachPsychScience.org   240
resistance and indifference to
        OER   150, 172
    author royalties   141, 151
    commercial textbook authors   151
    inappropriate adaptation
        concerns   141, 262
    intellectual property
        concerns   140, 187, 241, 259,
        262
    lack of content   152
    low levels of awareness   151
    risk-aversion   151
resistance to change
    see academic resistance to change.
Resources for the Teaching of
        Social Psychology
        website   240, 251
resources, open access to   53
review process
    BC Open Textbook Project   232
    Registered Reports   94
    TeachPsychScience.org   242
Rice University   10, 11, 149
risk-aversion   151
Roebuck, F.   51
ROER4D   76
Rogers, C.   51, 52
Rogers, E.   172, 274
rough consensus model   109
royalties   141, 151

S

safety concerns   121
Saylor Foundation   232
Sayre, Wallace   167
Scholarly Publishing and Academic
        Resources Coalition
        (SPARC)   16, 17, 149, 231

Scholarship of Teaching and Learning
        in Psychology   239
Science   15
scientific research   268
    cultural change   95
    incentives for increasing
        openness   94
    open science as default   269
    preregistration of analysis   95, 269
    reproducibility of   92, 93, 94
    sharing raw data   267, 269
    training to increase openness   94
    transparency in   91, 94
screencasts   126, 130
    and class cancellations   135
    assignments   135
    copyrighted material in   136
    creation   130
    distribution   131
    distribution in Open Education
        Commons   136
    streaming   131
    student responses to   131
    uses of   133
ScreenFlow software   130, 131
self-grading   121
self-questioning   52
self-verbalization   52
Seligman, M.   49
Seligman, Martin   190
service learning activities   54
SHARE project   95
sharing   140, 260
    see also research methods teaching,
        sharing ideas.
Shedd, John A.   163
Shockey, Nick   231
Shuttleworth Foundation   16, 202
Simon Fraser University   235
Siyavula   231
skillscommons.org   39
slide projectors   127
Social Psychology   94
Social Psychology Network   240

Society for the Teaching of
　　Psychology　239, 240, 243
Teaching of Psychology Idea
　　Exchange (ToPIX)　252, 253
software
　blogging　251, 253
　journal publishing　183
　open textbooks　120, 233
　presentation　126, 127
　screencast creation　130, 131
　see also open source software;
　　screencasts.
　web annotation tools　120
software access codes　156
source data publication　192
South African Institute for Distance
　　Education　14
SPARC
　see Scholarly Publishing and
　　Academic Resources
　　Coalition (SPARC).
Spellman, Bobbie　96
Spies, Jeff　92
Spohrer, James　10
Springer　182, 185
Srivastava, Sanjay　269
Stacey, Paul　231
standardized tests　49, 50
Stanford University　10, 11
STEAM (science, technology,
　　engineering, arts,
　　mathematics) projects　53, 54
Stommel, Jesse　117
streaming videos　131
student performance
　and commercial textbooks　81
student privacy　121
student-centered pedagogy
　　117, 145
student-created textbooks　120
student observers　52
student performance　70, 72, 75
　and open textbooks　172, 199, 271
student recruitment　72, 75

student retention　70, 75, 271
student satisfaction　72, 75
study aids
　see ancillary resources.
supportive feedback　143

T

Taylor, J. C.　107
Teacher Education in Sub-Sahara
　　Africa (TESSA)　18, 20
teacher training　20
teaching aids
　see ancillary resources.
teaching conferences　239
Teaching of Psychology　239, 246
Teaching of Psychology Idea
　　Exchange (ToPIX)　252, 253
teaching-oriented journals　239
teaching processes　50
TeachPsychScience.org　240
technology　125
　classroom technology　126
　GNU-Linux operating
　　system　103, 108
　podcasts　129
　see also open source software;
　　screencasts; software.
technology supplements, with
　　commercial textbooks　83
TED talks　74
TESS-India　18, 20
Textbook Assessment and Usage
　　Scale　81
textbooks
　see open textbooks.
Thanos, Kim　200
Thomas, F. C.　4
Thompson Rivers University
　　102, 110
Tidewater Community
　　College　149
TOP Guidelines　94, 269
ToPIX (Teaching of Psychology Idea
　　Exchange)　252, 253

Torvalds, Linus   108
Trade Adjustment Assistance
       Community College and
       Career Training Grant
       Program (TAACCCT)   39
traditional textbooks
   see commercial textbooks.
transformation of education and
       schooling   43
   assessment   48
   desired outcomes   48
   focus   48
   learning tasks   52
   openness   45
   organizational structure   55
   Psychology Departments   55
   purpose   47
   resources   53
   teaching processes   50
   transparency   46
   types of education   44
   work environment   54
Transforming Education   49
transparency   5
   in education and schooling   46
   in funding proposals   106
   in scientific research   91, 94
   open review policy   193
Transparency and Openness
       Promotion (TOP)
       guidelines   94, 269
Tri-Council Open Access Policy,
       Canada   268
Twitter   121, 230, 250, 253
Type I and Type II errors
       166, 249

U

Ubiquity Press   269
unaffiliated academics, and open
       access publishing   190
United Nations Educational, Scientific
       and Cultural Organization
       (UNESCO)   12, 15, 16, 19, 39

United Nations Universal
       Declaration of Human
       Rights   3, 110
University of British
       Columbia   119
University of California-
       Riverside   93
University of California, Davis   272
University of Canterbury   112
University of Southern
       Queensland   102, 110
University of Virginia   91
US Department of Labor   34, 39
Uses of the University, The
       (Kerr)   165
US Supreme Court   30
Utah State University   10, 198

V

Vazire, Simine   269
Veblen, T.   4
Vest, Charles   11
Vic Davis Memorial Trust   190
Vimeo   136
Virginia Virtual Library
       (VIVA)   159
Virtual University of Pakistan   16

W

Washington State Board for
       Community and Technical
       Colleges   230
web annotation tools   120
Weebly for Education   137
Weller, Martin   258
West Virginia University   215
Wiki Education Foundation   118
WikiEducator   104, 105
Wikimedia Commons   136
Wikipedia   35
   assignments   118
   Good Article status   118
Wilbanks, John   267
Wiley, David   10, 12, 38, 231, 271

Williamson, Daniel   231
Willmot, P.,   119
Wistia   136
Woodward, Tom   158
WordPress   120, 233
work environment   54
World OER Congress, Paris
        16, 39
Wright, Erik Olin   119

Wu, Stephen   190
WYMIWYG (What You Measure Is
        What You Get)   49

Y

Yale University   10
Yap, John   228
Young, J. R.   135
YouTube   16, 74, 136